Evolution and the Emergent Self

EVOLUTION AND THE EMERGENT SELF

The Rise of Complexity and Behavioral Versatility in Nature

RAYMOND L. NEUBAUER

Illustrations by Xuan Yue

Columbia University Press / New York

Columbia University Press
Publishers Since 1893
New York   Chichester, West Sussex
cup.columbia.edu
Copyright © 2012 Columbia University Press
All rights reserved

Library of Congress Cataloging-in-Publication Data

Neubauer, Raymond L., 1942–
Evolution and the emergent self : the rise of complexity and behavioral versatility
in nature / Raymond L. Neubauer ; illustrations by Xuan Yue.
p. cm.
Includes bibliographical references and index.
ISBN 978-0-231-15070-5 (cloth : alk. paper)—ISBN 978-0-231-52168-0 (e-book)
1. Human evolution.   2. Human behavior.   3. Behavior evolution.   4. Social
evolution.   5. Animal behavior—Evolution.   6. Evolution (Biology)   I. Title.
GN281.N45 2012
599.93'8—dc22
2011012708

c 10 9 8 7 6 5 4 3 2 1

References to Internet Web sites (URLs) were accurate at the time of writing. Neither the
author nor Columbia University Press is responsible for URLs that may have expired or
changed since the manuscript was prepared.

To Eva,

with gratitude for the love

and encouragement over the years

Contents

Evolution and the Emergent Self

Overview

We stand at the verge of an unprecedented understanding of our place in nature. With the latest particle accelerator, we can probe energies that existed less than a microsecond after the Big Bang, and with our current telescopes, we can peer 13.2 billion light-years into a universe estimated to be about 13.7 billion years old.[1] We have now sequenced the human genome entirely and are rapidly sequencing other species so that we can put together a full evolutionary story of life on this planet.

The universe revealed by modern science is one of continual change. The cosmos continues to expand from the cosmic fireball that began our universe, galaxies are condensing within vast clouds of dust and gas, stars are turning in the eddies of the galaxies, and planets have now been found in orbit around many stars. Even the land under our feet is shifting continually through plate tectonics. Within this universe of change, life as we know it on this planet exists within fairly narrow limits, not much beyond the boiling and freezing points of water. To survive, life must build an inner world that stays adjusted to change around it, and it does so with two main information systems: genes and brains. With base pairs of DNA and synapses between neurons, life is able to record information and build up responses that allow it to survive fluctuations in the environment. Genes, of course, came first and remain the main way information is passed from one generation to the next in most species. Nervous systems did not arrive until the most recent 15 percent of the history of life on Earth, when animal body plans got under way in a period known as the *Cambrian explosion*. Nervous systems can respond more quickly to change than genes, and with the arrival of humans, language and culture greatly expanded the amount of information that could be transferred from one generation to the next. The unprecedented dominance and control our species now exercises on the planet are testimony to the power of information to cope with the forces of nature.

In this book, I propose to look at life through two paradigm strategies of using information. In the first, life builds short programs in genes, brains, or both that can develop quickly. They require little parental investment per organism and have a fairly limited repertoire of behavior, and many individuals may be expended to find the few best suited to survive in prevailing conditions. Alternatively, life builds programs of higher information content in genes and/or brains that have a wide range of behavioral responses to environmental change. These more complex programs tend to take longer to develop and require a high investment per individual by parents. In chapter 1, I describe a contest between these two systems in the way the immune system fights disease, and I expand these ideas with two different kinds of selection described in ecology. Chapter 2 contains a brief overview of the history of life in terms of homeostasis, the ability of information systems to build an inner world that buffers life against fluctuations in its external environment. In chapter 3, I give a more precise definition of information content in genes and brains in terms of information bits, and I describe an increase in both systems over evolutionary time.

Humans clearly have taken the high-information road, but in chapters 4 to 6 I show that we are only the latest of a number of species that use slow development and wide behavioral repertoires as a way of dealing with a world of change. I describe a variety of species from very different lineages and different habitats whose main common characteristic is high relative brain size. They share a surprisingly similar set of qualities: complex communication and social systems, tool use, and an ability to both innovate and imitate. All manipulate their environment in intricate ways, albeit through very different appendages: an arm, a flipper, or a trunk. I describe this constellation of qualities as an emergent self, the result of a nervous system that has a higher level of objectivity regarding self, others, and its environment. In chapter 7, I describe hierarchical circuits in the brain that might make these higher levels of objectivity possible.

In the last chapters of the book I consider the growth of information and complexity from a cosmic perspective. The universe began in a uniform plasma of high energy that quickly differentiated into subatomic particles and atoms of the simplest elements. Heavier elements were cooked in the nuclear fires of the first stars, which flung this material into space in their death throes. This, in turn, made the first rocky planets possible. In chapter 9, I discuss theories in physics that see unity in the increasing complexity of matter from the first elements to molecules and life. The field of physics now recognizes a special class of "dissipative structures" that arise at energy gradients and use the flow of energy to "self-organize" structural complexity. Hurricanes and whirlpools are of this nature; they are transient forms that persist because of a constant flow of material through them, and they are able to maintain themselves as long as they have an energy gradient to feed on. All life requires nourishment to sustain its structure and goes through cycles of birth, maturity, and death, and thus may be a special form of the dissipative structures found all over nature. The theories of Eric Chaisson show a common pattern of rising energy flows that support increasing complexity of organization in both living and nonliving forms, and that thereby reveal a unity between the inorganic and organic realms. Chapter 9, on energy flows, covers Chaisson's ideas in relation to the concepts of complexity and entropy.

Life and intelligence have happened on this planet. How unlikely are we? There are many reasons to think that the appearance of life is an entirely natural process, likely to occur in the right set of conditions. Life may even be one of the natural states of matter, a self-organizing structure that appears at energy gradients with the right set of raw materials. We are still far from understanding the full complexity of the origin of life, but chapter 10 covers modern research that has traced many of the basic components of cells back to simple beginnings. Life makes use of the most abundant elements seeded into space by stars at the end of their lifetimes.

Conditions for the appearance of life may also be widespread across the universe. By February 2011, planet hunters had located about 525 planets outside our solar system. Then came the announcement that Kepler, the latest space telescope launched by the National Aeronautics and Space Administration (NASA), had located 1,235 more candidate planets, 356 of which are twice Earth-size or smaller. Fifty-four of these are in the habitable zone around their star.[2] James Kasting, a prominent researcher at NASA and a professor of geosciences, calculates in a recent book that the number of Earth-like planets in our Milky Way galaxy could be at least 4 billion, and there are some 80 billion other galaxies out there.[3] This does not necessarily mean, however, that ET is likely to show up soon, or even that we can expect to find signals easily from other inhabited worlds. Kasting also calculates that planets with advanced civilizations, if they exist, would be separated by an average of 250 light-years. This makes long-distance conversations, to say nothing of interstellar visits, highly problematic. Chapter 11 deals with the prospects for habitable worlds around other stars.

For much of the twentieth century and its surrounding eras, the spirit of science often seemed to be dehumanizing. It appeared to emphasize an impersonal world of swirling atoms that had no connection to human yearnings and hopes. The theme of alienation was sounded early during the Industrial Revolution by Matthew Arnold's "Dover Beach" (1867), with its depiction of a world that "Hath really neither joy, nor love, nor light, / Nor certitude, nor peace, nor help for pain."[4]

This mind-set was codified in modern times by existentialism in the work of people such as Jean-Paul Sartre and Albert Camus. But it is a strange notion, really, that humanity, a product of nature, should feel alienated from the process that produced it. Our current understanding of evolution is that behavior as much as physiology is molded by natural selection. Why then, should it produce a mind that is despairing and alienated from the process that created it? Are humans an aberration, or is our culture unnatural? No bees search for a nectar that does not exist. Why should people have a need for meaning that is not there?

The universe uncovered by the latest science may be a garden rather than an existential desert. Life may be springing up in solar systems scattered all over the cosmos. Intelligence may be a major theme of evolution wherever it has had time enough to play out its possibilities. As I will describe in the following chapters, a high-information pathway is one of the basic strategies of life, and it leads repeatedly to an emergent self: a consciousness that seeks to express itself in complex communication and social relations, with intricate manipulation of its environment. This may be a universe *for* life, where the lights that dot the night sky are like a field strewn with wildflowers. The evolutionary processes we find here on Earth may be common all over the cosmos, and for those so inclined, this makes a religious view of evolution also possible. The full implications of the Copernican principle, that there is nothing unusual about this corner of the universe, indicate that the evolutionary processes we have witnessed on this planet may be widespread all over the cosmos. In the final chapter I consider some of the philosophical implications of the wide-ranging view of nature presented here.

1

The Immune System: A Parable

Life has increased in information content over time and has also become more homeostatic; that is, it has built an "inner world" that is resilient to fluctuations in the external world. Life likely began in the ocean, where temperature variations are not as extreme as on land. Variations in salt content, pH, and of course water level are also not as extreme in the seas. Over time, life learned to buffer itself against these variations. When it came out of the water, it carried an internal circulatory system that still mirrors the salt content of the ocean. It evolved scales to guard against drying out on land. Reproduction became more homeostatic: fish and amphibian eggs cannot leave the water, but reptiles evolved the amniotic egg that is surrounded by a hard shell and has multiple internal membranes to protect the embryo. Mammals went a step further: They retain the egg within the mother, and she has become its main source of nourishment and protection.

This increase in homeostasis mirrors the increase in information content of organisms. We may think of genes and brains as accumulators of programs to find more ways to respond to changes in the environment. The more alternatives an organism has in response to challenges, the greater its chances of survival.

Life builds an inner world that must stay adjusted to changes in the outer world. It tracks changes in the environment with two fundamental strategies that can be seen as end points of a continuum. It either builds many organisms of low information content that can develop quickly, in order to find the variants best suited to fit new conditions, or it builds versatility into single individuals. The latter have high information content in genes and/or brains and possess wide behavioral repertoires that can adjust to change. The first strategy expends many individuals to find the few best suited to meet new conditions, whereas the latter invests resources in the long development of fewer individuals that have broad behavioral repertoires.

Humans represent one extreme in the strategy of long development and behavioral flexibility, but this strategy has evolved repeatedly in different lineages: in dolphins and elephants, among other mammals; in some songbirds and parrots, among birds; and in squids and octopi, among invertebrates. All these organisms have high information content in genes and/or brains, and they share qualities in common of complex communication and social systems, with an emphasis on learning and behavioral flexibility as strategies of survival. Humans are the latest and most extreme example of a strategy that life has used repeatedly to survive the fluctuating challenges of the environment.

The Immune System

The way that life uses systems of both high and low information content can be seen in how the immune system battles disease, and because this competition also illustrates some fundamental principles about evolution, I will explore this system in some depth. I will not try to cover the whole immune system, which would take a long book in itself, but rather one important wing that illustrates principles that will be used repeatedly in this book.

The battle between microbes and the immune system is between two systems of very different information content. Humans have about 22,300 genes, whereas viruses vary from a few genes to a couple of hundred. Human immunodeficiency virus (HIV), the virus that causes acquired immune deficiency syndrome (AIDS), does its tricks with only nine genes. Most bacteria have anywhere from about 350 to 8,000 genes. This difference in size is reflected in generation times. An average bacterium can grow and divide in 20 to 30 minutes, whereas an average animal cell needs at least a day. These differences are also reflected in mutation rates of a population. *E. coli* is a common bacterium that lives in the human lower intestine and has about 4,300 genes. Each day, some 20 billion new bacteria are born[1] (many are expelled with our feces) and of these, about 9 million will carry a new gene mutation. A virus like HIV has a mutation rate 200 times higher than this.

This ability to multiply and change would overwhelm the immune system except for a unique genetic mechanism used only here in our bodies. The immune system is able to mix and match pieces of its genes to create unique combinations to fight disease. This process is seen most clearly in the production of antibodies.

Antibodies are Y-shaped proteins produced by B cells, immune cells that develop in our bone marrow and are then released into our circulatory system. Antibodies fight microbes in a variety of ways. The two upper ends of the Y form very precise "grabbing" shapes that can attach to molecules on the surface of microbes and thereby inhibit their activities (figure 1.1). Pathogens covered with antibodies are also targets for macrophages (literally, "big eaters"), immune cells that engulf antibody-covered microbes and digest them.

Each B cell, as it matures in the bone marrow, becomes a specialist for only one shape of antibody. Yet we produce millions of different antibodies in such variety that it is almost certain that no matter what microbe gets into you, you will have some B cells that make antibodies that can latch on to it. For example, the human race presumably did not see the AIDS virus before the first half of the twentieth century. Yet the body produces such a variety of B cells that should that virus enter your system, you will have some B cells with antibodies that can recognize it and attack.

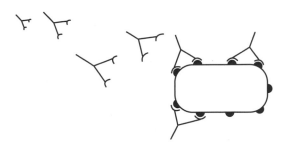

Figure 1.1 A microbe being covered by Y-shaped antibodies.
The molecular shape on the ends of this type of antibody fit shapes on the surface of the microbe.

A selection process has to take place for this system to work. When an infection starts in the body and a microbe begins to multiply, a great variety of B cells come into contact with the invader. Each B cell is a specialist for only one shape of antibody on its surface, and most B cells will not produce an antibody that can attach to a given microbe. Only a small minority of the B cells will have the right shape that fits. Thus, at the beginning of an infection, the body searches, as if through a deck of cards, for the right one to play in this particular situation.

Having found the right card, it now multiplies it. That is, if a B cell is displaying an antibody that can attach to the foreign microbe, the act of docking with the microbe stimulates that B cell to reproduce. It now multiplies in great number as a clone of itself (many copies of that particular B cell). This large clone then separates into two populations. One group secretes its particular antibody in large numbers, and these antibodies fight the pathogen in the ways just described. Another group remains as a regiment of "memory cells." They continue only displaying their antibody, but are ready to pounce should that microbe ever show up in the body a second time.

This process is known in immunology as *clonal selection*. You first go through a selection process. Of the huge number of B cells turned out from the bone marrow, only a small number have shapes that can fit the invader. The selected cells are then cloned; that is, they copy themselves in large numbers.

The immune system produces millions of different B cells, yet there are only about 22,300 human genes. How can it produce all this variety? The immune system relies on a unique genetic mechanism found nowhere else in our bodies. The genes for antibodies come in sections and the immune system stitches them together the way quilts are made of patches. Each section of the gene codes for only part of the grabbing end of the antibody, and there are dozens of variants for each section so that the diversity of all the possible combinations is enormous. As a B cell matures, it chooses each section at random so that no two B cells are likely to begin life displaying the same shape of antibody. The process is random because it is not directed toward any goal, and one outcome is as likely as another. Most of the resulting B cells will be useless and will gradually die away, but the few that have shapes that allow them to latch onto the pathogen will be stimulated to form a large clone of themselves.

Microbes are using their own random number generator. Mutation, as far as we know, occurs entirely at random. Base pairs of genetic material may change because of errors by the enzymes that copy genes or because of radiation or carcinogens in the environment. The high mutation rates and short generation times of microbes allow them to turn out a large number of variants in a short time, as already described. Many of these variants will probably be less infectious than the original microbe. But some of the variants may hit on combinations that can elude both the immune system and any medicines we apply, and that is one of the main reasons a disease like AIDS is so hard to combat (HIV also kills the very immune cells meant to hunt it down, like a terrorist who not only continually changes his disguise but also shoots the police trying to track him).

Over the long term, an arms race is going on between pathogens and the immune system. The microbes are seeking new combinations to attack the host, and the immune system is fighting back with new combinations that can destroy invaders. Each is using a random process that produces a great deal of variation because they cannot anticipate the moves of the other. The immune system cannot have foreknowledge of what pathogens it will encounter, and there is no "mind" within pathogens that can forecast either our medicines or the versatility of our immune defenses.

But there is a difference in the value of the individual in this contest. Microbes sacrifice many variants, seeking the forms best able to infect the host. The immune system builds versatility into a single being. The system takes longer to act because it has a complex series of steps, but it is designed to save the life of a single individual. One system has a relatively simple genetic program and short generation times and can afford to sacrifice many variants in the hunt for the best ones. The other has a complex genetic program, longer development time, and builds versatility of response into single individuals.

This difference in generation time also explains the typical time course of a disease. In the first few days of an illness you usually feel lousy, as the generative power of the microbes outruns the speed of your immune system. Only after a complex selection process can the immune system go into high gear to fight the disease.

We can see here two paradigm strategies of life: simple and quick versus complex and slow in development—the hare and the tortoise, only in this case the slower one takes longer to reach its goal because it is going through a more complex series of steps.

The race takes place on the fields of infection, between simple pathogens and the immune systems meant to combat them, but it also distinguishes between whole groups of organisms that can multiply quickly and take over an environment and those that have slower, steadier growth.

These strategies are two end points of a continuum, with some organisms having combinations in between. They are two ways of solving the fundamental need of life to meet the challenges of an ever-changing environment.

Randomness and Meaning

Both the immune system and microbes use a random number generator to produce variation, but this does not mean that life is simply random. In a universe of change, this is probably the only way life could be. Either we live in a static universe with fixed species that were created with all the means they need to survive, or we live in a changing universe that must have mechanisms of adjustment.

What kind of universe has been revealed to us in the past century? The galaxies are flying outward from the explosion of the Big Bang, stars are igniting and moving within the galaxies, the continents are shifting under our feet, and climate is continually changing around us. Our best supercomputers can barely forecast the weather for a week. How then can any organic system really predict what tomorrow will bring?

Life has responded to this challenge by hedging its bets, by using a random number generator to produce many possibilities. Although the raw material is random, the results are not. Not just any virus can infect us, only ones with combinations to unlatch the gateways to our cells. The immune system answers this challenge with a random process, but it is randomness used for a purpose. Like a computer searching many possibilities to crack a code, it seeks the combinations that can best neutralize the threat of a pathogen.

Complexity and the Individual

The battle between the immune system and disease illustrates two paradigm life strategies found repeatedly in nature. They represent two different kinds of investment of resources by parents. Simpler organisms, with shorter programs, tend to produce young in great numbers. They develop quickly and need relatively little investment per offspring by parents. Many of them may die, but some will have variations that fit the new conditions, and they will most likely be the parents of the next generation.

More complex organisms, with larger programs, will probably take longer to develop for the basic reason that it takes longer to build complex things than simpler things. Such organisms may require a lot of parental investment, and therefore the parents will likely have fewer young. However, behavioral versatility will be built into these offspring. The variation will be in a single individual, in the versatility of its responses to challenges from the environment.

These two types of species are sometimes called *opportunistic* and *equilibrial* in ecology. Simpler, faster organisms can take over an environment by multiplying quickly, like a microbe that has invaded the body. Its population may just as quickly crash when conditions change. Equilibrium species have steadier population numbers; they have a behavioral versatility that lets them adjust to changing conditions.

Equilibrium species also tend to be more homeostatic, having more internal systems that are buffered against fluctuations in the environment. For example, we may consider a "warm-blooded" mammal more homeostatic than a "cold-blooded" reptile. The mammal's body temperature does not fluctuate as much with external temperatures, and so it can be active in a larger variety of conditions than a reptile.

Are equilibrium species better than opportunistic species, or are all opportunistic species seeking to become equilibrial? Notice that they are two ways of dealing with variation in the environment, two ways of tracking change. Sometimes one strategy is more successful, and sometimes the other. When one talks of long-term trends in evolution, it does not mean that all species are seeking to become one or the other.

Two Kinds of Selection

In ecology, these two paradigm life strategies are sometimes seen as the products of r- and K-selection, two terms that are used to describe the way populations grow.

A typical growth curve for a population is shown in figure 1.2. Starting from small numbers, a population may have a period of rapid increase (r), until it reaches the carrying capacity (K) of its environment. The carrying capacity, or maximum possible population size, may be

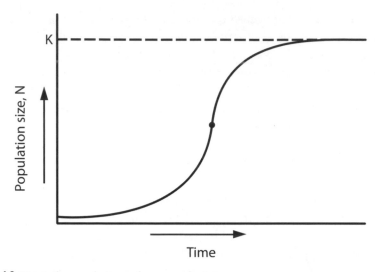

Figure 1.2 Typical growth curve for a population.
The point in the middle of the curve is where r, the rate of increase of the population, is at its maximum. Growth will level out at K, the carrying capacity of the environment. (From Neil A. Campbell and Jane B. Reece, *Biology*, 6th ed. [San Francisco: Benjamin Cummings, 2002], 1162. Adapted by permission of the publisher)

due to many things: the large population has outrun its food supply, or accumulating wastes are poisoning its environment, or the large population may attract predators or disease. No population can increase without limit, and there will come a time when the rate of increase (r) is balanced by the rate of death, and population size levels out to the maximum carrying capacity (K) that its environment can sustain.

Fitness in evolutionary theory is determined by the number of offspring an individual leaves in the next generation, whether they be his own, those of relatives, or both. The individual in a population who leaves the most offspring that survive to reproductive age has the highest fitness. Should not all species seek to breed continually, leaving as many offspring as possible? The answer may depend on the environment. There are situations where opportunistic species, which can reproduce and grow quickly, will have an advantage. When a new habitat opens up suddenly and unpredictably—for example, after a fire or flood—plants that produce many seeds that disperse easily may be able to colonize that area quickly. In contrast, a well-established forest with many tall trees may rarely have open spaces for new plants. Here large individuals, with long life spans, may best compete for the available light and nutrients.

Different species have different survivorship curves. A scallop, for example, can release millions of eggs, only a few of which will find a substrate to grow on and thus mature.[2] The young of such organisms have a very high mortality rate, but if they make it past a hazardous initial period, survivorship to adulthood is likely (see the type III survivorship curve in figure 1.3). Other groups of organisms, such as mammals (especially humans) have type I survivorship, with much lower death rates for their young and a high proportion of mortality in old age. Other species have fairly steady mortality at any age (type II), and some species have combinations of these three curves.

Parental investment usually differs according to the survivorship that a species can expect. Scallops produce millions of eggs that do not have much yolk and that are randomly dispersed by ocean currents with the expectation that a few will find a good place to settle and grow. It does not make sense for a parent to invest a lot of nutritious yolk or care into any one egg. Humans, on the other hand, as well as many mammals and

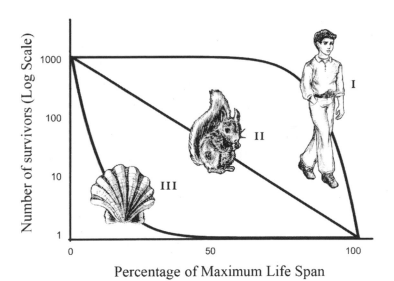

Figure 1.3 Survivorship curves for three different types of organisms.
Scallops have a very high mortality rate as larvae (type III), whereas humans in developed countries usually have a high survival rate until old age (type I). Squirrels have constant mortality at any age (type II). (From Neil A. Campbell and Jane B. Reece, *Biology*, 6th ed. [San Francisco: Benjamin Cummings, 2002], 1155. Adapted by permission of the publisher)

Table 1.1. Some of the Correlates of r- and K-Selection

	r-selection	K-selection
Climate	Variable and unpredictable	Fairly constant or predictable
Mortality	Often catastrophic	Steadier rate
Survivorship	Often Type III	Usually Type I or II
Population size	Variable in time, nonequilibrium; usually well below carrying capacity of environment	Fairly constant in time, equilibrium; at or near carrying capacity of the environment
Competition	Variable, often lax	Usually keen
Selection favors	1. Rapid development 2. High maximal rate of population increase 3. Early reproduction 4. Small body size 5. Single reproduction 6. Many small offspring	1. Slower development 2. Greater competitive ability 3. Delayed reproduction 4. Larger body size 5. Repeated reproduction 6. Fewer, larger progeny
Length of life	Short, usually less than a year	Longer, usually more than a year

SOURCE: Adapted from Eric R. Pianka, *Evolutionary Ecology*, 6th ed. (San Francisco: Benjamin Cummings, 2000).

birds, lavish a great deal of individual care on each offspring to help ensure its survival.

The characteristics of these two strategies are summarized in table 1.1. It describes two suites of characteristics that are often found in species that may be called *opportunistic* versus *equilibrial*, or *r-selected* versus *K-selected*. Opportunistic, or r-selected species, tend to have rapid rates of increase, small size, many offspring, rapid development, and little parental care. They are able to colonize variable or unpredictable habitats quickly but may also experience catastrophic mortality when conditions change.

Equilibrial, or K-selected, species have fewer young, with slower development, and a lot of parental care. They often exist in more constant or predictable environments where competition is keen and long-term survival skills, in terms of either behavioral versatility or physical growth, are important. Physical growth may involve simply growing to large size, or developing defensive weapons like antlers or complex toxins. Population size in such species often stays at a more steady level, without the boom and bust of r-selected species.

These are broad generalizations that describe two end points of a continuum. Many species have a mixture of these characteristics, and some are not well described by these categories at all. Some biologists believe that these definitions are too broad, involve too many variables, and are therefore hard to test scientifically.[3] Nevertheless, support for these concepts comes from a variety of sources that indicate some fundamental distinctions among both animals and plants.

Plants that live for only a year (annuals) produce a large number of seeds with relatively small amounts of nutrients but high dispersal ability, such as dandelions.[4] Annuals can quickly colonize fleeting areas, such as bare spots in your lawn or sunny patches in the forest before the trees have put out leaves in the spring.

The tall grass-like plants that grow on coastal dunes have windblown seeds that germinate, grow quickly, and then produce many more seeds.[5] They put almost no resources into antiherbivore defenses or drought adaptations. They quickly colonize new areas of sand that periodic ocean storms make available.

In contrast, a K-selected area for plants would be a tall forest where almost every niche is filled and plants are competing intensely for light.[6] Adaptations for tall growth, using soil poor in nutrients, or synthesizing chemical compounds that inhibit predators or disease will be an advantage. Growth and reproduction are comparatively slow. Examples are forests of long-lived conifers such as redwoods, firs, and bristle-cone pines.

Producing chemicals that inhibit predators can be expensive for plants and may slow overall growth. Biologist Geerat Vermeij notes that slow-growing land plants and marine algae often have high concentrations of tannins and phenols, compounds that are distasteful to herbivores.[7] In contrast, fast-growing plants usually make fewer noxious chemicals and have a high turnover of photosynthesizing tissues. They are attacked more heavily by grazers, but can replace their leaves quickly.

The *fast-slow continuum*, as it is called in ecology, has found much support in different areas. It is a version of "live fast, die young" and is a strategy used by many organisms that experience a high mortality rate in early adulthood. A species will invest most in that age class that contributes most to reproduction. If it can expect high mortality early on, it will get reproduction over with as quickly as possible, perhaps in one "big bang" of group mating.

Humans lie at the other end of the continuum, with extensive parental care, delayed reproduction, and long life span. This pattern is found in a number of animal species that have high relative brain size or use extensive learning as a primary strategy for survival. A study of 12 orders of mammals found a significant relationship between maximum life span and brain size.[8] This pattern applies to both living and fossil forms.[9] The larger the brain in relation to body size, the greater the longevity of the species.

The high metabolic demands of a large brain may slow growth patterns overall. Primates, carnivores, and elephants all have long gestation periods, late weaning age, and large relative brain size.[10] All these groups have an extended juvenile period during which learning and play are prominent. In birds also, delayed maturation and large brain size are related to long periods of learning, either singing for songbirds or hunting skills for raptors. Parents invest in smaller broods, with attendant delayed maturity, longer reproductive cycles, and longer lives, but also with complex behavioral skills for survival. A slowly maturing brain appears to be a pacesetter for the development of the body in general.[11]

Brains as well as genes can store information, and I suggest information content may be the defining quality that distinguishes the two paradigm life strategies in opportunistic and equilibrium species. Organisms of high information content in genes and/or brains take longer to develop for the basic reason that complex things take longer to build than simple things. They have versatility of response built into the individual and can deal with a variety of fluctuations in the environment. They possess many of the characteristics associated with K-selection: small litters with slow development of the young, allowing a long period for learning, and fairly steady population numbers because individuals have a variety of skills to cope with change.

Organisms of lower information content in genes and/or brains tend to develop more quickly. There is less parental investment per individual, but more are produced. They have shorter, less flexible programs, and many individuals may be sacrificed to find the combinations best suited to match new conditions. These individuals can quickly take over an environment so that populations go through characteristic "boom and bust" cycles.

Humans rely on an extreme form of a strategy found repeatedly in nature: the accumulation of information to increase the versatility of response to a changing environment. Programs are encoded not only in genes and brains but in languages, books, and computers. In this sense, human culture is not an anomaly but an extension of K-selected strategies found repeatedly in nature. It might even be suggested that a species like us is inevitable if the evolutionary process is given enough time to run. All of nature is not laboring toward greater complexity, for we saw that short programs with quick development are an alternative way of dealing with change. But it may not be too far-fetched to suggest that, given the right conditions and enough time, a species like us is in the cards and the seeds of consciousness are planted within the evolutionary process.

2

Voyages into Homeostasis

Homeostasis—a steady internal state that is buffered against fluctuations in the environment—is a far-reaching concept in biology. We have numerous homeostatic mechanisms in our bodies to keep just the right levels of sugar, salts, calcium, oxygen, and other vital molecules in our blood. Homeostasis can also be thought of as a behavioral quality: having a large variety of responses to changes in the environment. An ample repertoire of responses allows an organism to maintain its activities despite changes in its surroundings. Warm-blooded animals (homeotherms) like mammals and birds, for example, can remain active in winter when cold-blooded animals like insects and reptiles have to shut down.

Humans stand at a peak of both information gathering and homeostasis. These two concepts are related to each other. Information lets our species combat disease, secure raw materials, and improve its crops. In industrialized countries, humans live in cities where the lights never go out, rooms are kept comfortable regardless of the weather, and food and energy are in plentiful supply.

In a world of change, homeostasis is one of the major themes of life. Species seek to stabilize and expand conditions that are optimal for their way of life. Behavioral versatility is only one of many strategies life uses to respond to an environment of constant change. In this chapter I will give a brief overview of the history of life to show the drive toward homeostasis in both the animal and plant realms. The human strategy of collecting information as a way to deal with fluctuations in the environment can be seen as an extension of the drive toward greater homeostasis that has guided life from its beginning.

Life began in water, where the range of conditions is less extreme than on land. Temperatures do not shift as widely as in the air, and the effects of storms and winds are greatly moderated a few feet under water. Ocean water has dissolved salts that are suitable for the existence of life, and its buoyancy did not require much in terms of skeletal support in early life forms. Only slowly did life acquire the ability to venture from the nurturing embrace of the sea.

Life began as single cells and remained so for most of its duration on Earth. Only some 600 million years ago (mya)—a mere 15 percent of the time life has existed on our planet—did the experiment of multicellularity really get under way. Single cells are closer to their environment: a mere membrane and perhaps a cell wall separate them from nutrients, gases, and water. One of the main reasons cells remain small is that simple diffusion can take care of the exchange of nutrients and wastes between the environment and the cell's interior.

The earliest life forms were barely separated from their environment. Sponges, which are probably closest to the ancestral forms that led to animal life, have relatively unspecialized cells that can take in nutrients and release wastes directly into seawater. Water passes freely through the many pores in the sponge's body. Its organization is so loose that its cells can be passed through a sieve and they will reaggregate as a functional sponge on the other side.

Jellyfish (*Cnidarians*) are also near the base of the evolutionary tree of animals and probably have changed little in hundreds of millions of years. They are made of only two cell layers in which each cell is in close proximity to water. They have a relatively unspecialized digestive cavity with a single opening that functions for both ingestion and excretion.

Later animal evolution devised three cell layers and more complex organ systems. An internal circulatory system evolved that carried its own fluid, distributed nutrients, and took up wastes from cells within the body. This allowed a great deal of specialization of individual organs and the construction of more complex body plans.

This internal environment could be maintained at near-optimum conditions for the life of cells and was buffered from changes in external conditions. Our own bodies have numerous mechanisms to keep circulatory fluids at optimum levels for life. Sugar concentration, for example, is precisely monitored. If it is too high, insulin causes it to be stored away, and when it gets too low, sugar is taken from storage and returned to the blood.

The specialization of organs also required the transmission of information. Because different groups of cells had different roles in maintaining homeostasis, they needed to be able to communicate with each other. Chemical messengers like hormones, and, in animals, a nervous system became necessary. Specialization, homeostasis, and the sharing of information are all involved with each other.

The homeostasis of reproduction for both animals and plants has also increased. It began with the liberation of eggs and sperm from dependence on water to meet each other, as well as the evolution of multiple internal layers to protect an embryo. In the history of life, reproduction became increasingly buffered from the environment. The eggs of most fish and amphibians have a thin, gelatinous coat. They generally do not have much yolk and are relatively small and numerous, and the young hatch more quickly than is found in the development times of many land vertebrates. The egg's thin, porous coat requires that it remain in water to prevent drying out.

Although amphibians made the transition onto land, they must still remain near water, not only for reproduction but to keep their skin moist. They are subject to drying out, and because they do some of their breathing through their skin, it must have a thin film of water to be able to capture oxygen.

Reptiles made a number of innovations that truly liberated vertebrates from a watery medium for the first time. They developed tough, keratinized scales that protect their bodies from water loss and death. Even more remarkable is the amniotic egg that protects the embryo within a hard calcified shell while surrounding it with fluid-filled sacs that buffer it from external disturbances and provide an internal watery environment. Plentiful yolk can be supplied to nurture its development, and a sac is provided to hold wastes until it hatches. Reptiles also developed internal fertilization so that sperm could be directly deposited with a particular female and not be subject to the vagaries of gametes shed into ocean currents.

Mammals took homeostasis several steps further. The mother's body is now protection for the embryo and its source of nutrients. It is further buffered from changes in the environment as well as protected by the survival skills of the adult. After birth there may also be a period of instruction during which the juvenile learns its foraging skills.

There is a progression in homeostasis during different stages of mammalian evolution. The early-evolving monotremes still lay eggs but provide lactation for hatchlings. Marsupial eggs still have a thin shell inside the mother and are nourished by both egg proteins and maternal nutrients.[1] There is a short gestation period, and the young emerge in a highly immature form. They then crawl into the mother's pouch, where they receive nourishment from her teats.

Placentals were the last order of mammals to appear, and they dispense with a long-lasting egg shell. The embryo completes much more of its development inside the uterus, and a unique organ, the placenta, allows fetal blood capillaries to travel near maternal ones, where an exchange of nutrients and wastes takes place. The mother breathes and eats for the embryo, and even her immune system protects it from pathogens. Evolution has therefore gone beyond mere liberation from a watery environment and supplied additional layers of protection for the young. Keeping young inside the body also helps with temperature homeostasis. Instead of having to incubate eggs, as birds still do, the mother herself provides a warm environment for development.

William Cannon, who pioneered ideas of bodily homeostasis with his book *The Wisdom of the Body* (1932), also understood the concept in evolutionary and social terms. He wrote that reptiles and amphibia were less homeostatic "in the sense of having a controlled internal environment which liberates them from the vicissitudes of the external environment."[2] In terms of human society, he viewed an advanced culture as one that frees people from adversities such as hunger, drought, and disease.[3]

Plant Homeostasis

Plants also evolved a variety of homeostatic mechanisms when they conquered the land. Evidence indicates that plants also originated in the sea and that green alga is the probable ancestor of all land plants. Both groups use the same chlorophylls for photosynthesis, put cellulose into their cell walls, and share a variety of gene sequences. Some green algae form mats at the edges of drying shorelines, making what was probably the first environment that led to the transition onto land.

Mosses are the most familiar plants of a group known as the bryophytes, which were probably the first to colonize the land. They are all short plants, as they have not developed a good means of support. The buoyancy of water can support tall structures such as ocean kelp while plants remain in water, but on land, new means of bearing up under gravity must be found. Bryophytes have little or no conducting vessels for distributing water or food; nutrients move mainly by diffusion. They also lack true roots or leaves and depend on a film of water to allow sperm to swim across from one plant to the eggs of another. After fertilization, an embryo begins growing from the tissue of the parent and receives nutrients from it.

Vascular plants, which evolved next, are more homeostatic in a number of ways. They can retain water within conducting vessels of their tissues and sink roots into the ground, where water is in more steady supply. They are therefore capable of more continuous growth than the bryophytes. Although the bryophytes have photosynthetic cells in their stems, the vascular plants evolved the first true leaves.[4] The seedless vascular plants also evolved a new cell type, the tracheid, which can join end-to-end to form tiny tubes of a vascular system that transports water around the plant body. These cells could also be reinforced with lignin that provides strong support in cell walls. This allowed the evolution of much taller land plants than the Earth had seen before. But reproduction in these plants still requires a film of water to allow sperm to swim across to eggs.

Forests of seedless vascular plants dominated the landscape of the later Paleozoic era, 360 to 290 mya. During that time, Europe and North America were located near the equator and covered with warm, shallow swamps. Some of these plants grew more than 36 meters (120 feet) tall, and their accumulated debris formed the large coal deposits found around the world today.

Growing among the seedless plants of the later Paleozoic were the first seed plants (gymnosperms), which rose to prominence at the end of the age as the swamps began to dry up and the climate became cooler. Gymnosperms are represented today by conifers such as pines, firs, and redwoods, and by ancient forms such as cycads and ginkgos.[5] During most of the Mesozoic era, the age of dinosaurs, gymnosperms were the dominant land plants. Like the reptiles, their reproduction became fully independent of water for the first time. Unlike the earlier evolving mosses and ferns, which still require a film of water for sperm to swim across to eggs, gymnosperms have pollen that can be carried on the wind over long distances. Pollen is actually a tiny organism (the gametophyte) that germinates and produces sperm only after it lands on the female ovule. When the seed develops, it contains a tiny embryo, as well as nutrients supplied by maternal tissue and a hard seed coat. Fertilization may occur more than a year after pollination, and the seed can lie dormant for a long time until the right conditions initiate germination of the next generation.

The embryos of nonseed plants develop directly into adults and must live or die, depending on local conditions.[6] But gymnosperms are more homeostatic: they do not need a film of water for sperm to meet egg, and they can delay both fertilization and germination until an optimal time. The seed coat also protects the embryo from the harsh effects of ultraviolet light.

Flowering plants (angiosperms) first appeared in the Mesozoic and began to really diversify near the end of that era. They came to dominate the land and displaced the gymnosperms in most locales in the late Mesozoic and early Cenozoic (the age of mammals). Like mammals, they took homeostasis one step further. Gymnosperm means "naked seed," and the group is named for the exposed condition of its seeds on cones or specialized leaves. Angiosperms, however, sequester their eggs deep inside an ovary, where it is protected from predation by insects and herbivores. They also evolved a unique tissue to nourish the embryo, the endosperm, which is especially high in nutrients because it is produced by cells that have three complete sets of chromosomes (3n). Like the gymnosperms, their seeds can lie dormant for a long time until the right conditions foster sprouting.

The world grew more deeply into relationship with the evolution of mammals and flowering plants. Mammals have more individualized relationships with each other, especially in the long-term care that mothers provide their young. Flowers also have more individualized relationships with each other. In most flowers, pollen is carried from one plant to another by an insect or animal vector, instead of being blown haphazardly on the wind, as with gymnosperms.[7] An alliance has been forged between the animal and plant realms. Some flowers can be pollinated only by a single species of animal, and many angiosperms grow fruit to lure animals to eat and disperse the seeds packed inside. They take advantage of the mobility of animals to carry seeds to new areas, and animal droppings serve as useful fertilizer.

The two realms of life make each other possible in some ways.[8] Mammals and birds have very high metabolic needs because of the activities of being warm-blooded (endothermic). The high nutrient values of endosperm and fruits supply much of that food. Humans have settled on angiosperms for almost all their domesticated crops: rice, wheat, corn, and potatoes. Plants, in turn, benefit from their alliance with animals in terms of pollination and the dispersal of seeds.

Two great crises contributed to the domination first of the reptiles and gymnosperms and then of the mammals and angiosperms. Near the end of the Paleozoic, all the landmasses came together to form a supercontinent, Pangea (literally, *whole world* [figure 2.1]). Inland coastal seas and swamps were drained, and the continental interior became drier and its weather more variable. Large volcanic eruptions occurred in what is now Siberia.[9] Under conditions like this, organisms that were less reliant on water and could survive in arid inland areas had an advantage, and the reptiles and gymnosperms became dominant.

Another great transition took place at the end of the Mesozoic.[10] Many biologists believe the climate was becoming cooler at that time, there were changes in sea levels, and perhaps large volcanic eruptions occurred in what is now India. These changes culminated in the impact of a large asteroid about 65 mya just off the Yucatan Peninsula in Mexico. A worldwide underground layer of iridium, an element rare on Earth but common in asteroids, dates to this period and testifies to the global nature of the catastrophe.

There was a mass extinction at the end of the Mesozoic, as there had also been near the end of the Paleozoic. It has been argued that the species that survived were purely a matter of luck and that without the asteroid impact the dominant animals today would be very different. This may be so, but notice that after each crisis it was the more homeostatic multicellular organisms that flourished. Those that were more independent of changes in their surroundings and could grow under a greater variety of conditions came to dominance. In the drier conditions of the Mesozoic, gymnosperms and reptiles had an advantage over earlier groups that were more dependent on water. Flowering plants were already succeeding gymnosperms in the later Mesozoic, and the Cenozoic only hastened that transition. Mammals remained small during the age of the dinosaurs, when steady warm conditions of the Mesozoic probably would not have given them an advantage over large-bodied reptiles. But in the cold, variable conditions of the late Mesozoic and Cenozoic, they had an advantage, especially after the demise of the large dinosaur predators.

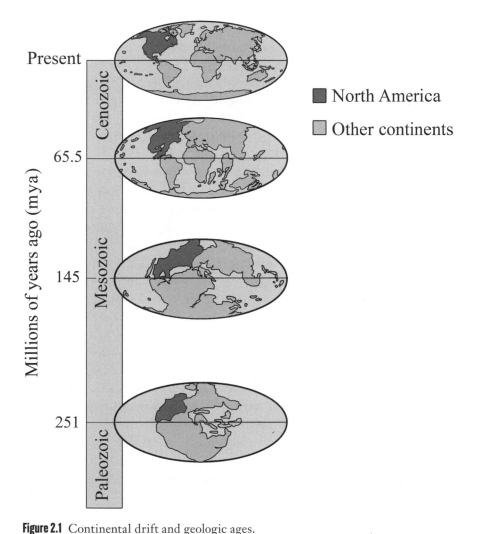

Figure 2.1 Continental drift and geologic ages.
At the beginning of the Mesozoic, all of Earth's landmasses formed the supercontinent Pangea. The landmass that would become North America is shaded for orientation to the changes in the continents. (From Neil A. Campbell and Jane B. Reece, *Biology*, 6th ed. [San Francisco: Benjamin Cummings, 2002], 490. Adapted by permission of the publisher)

There are indications that some dinosaur species were evolving toward larger brain size, advanced parental care, and homeothermy, although the evidence remains controversial.[11] But we do not need to look that far to see a trend. The bone structure of birds is so close to that of theropod dinosaurs that some biologists say the dinosaurs are not extinct, they are singing in the trees. Birds have also taken the pathway to greater homeostasis. They are warm-blooded, and many avian species take extensive care of their young.

Probability predicts that more different events will happen over time. On a world with active geology, in a solar system that condensed from a cloud of dust and gas, over time there will likely be more debris from outer space, more continental drift, more changes in climate and sea levels. For life that exists within a relatively narrow range of conditions, there must be a strong tendency toward homeostasis, toward building an inner world that is buffered against fluctuations in the outer world. The changes described here are probably not just accidental (although mutations supplied the raw material). Those groups that came to dominance in each era evolved new ways to be independent of their surroundings. They did so by building up alternative programs, different ways of responding to changing conditions.

This is not to suggest that all organisms are somehow striving to become more homeostatic. Rather, in the range of possible adaptations, behavioral flexibility can provide advantages in changing environments. This is a hallmark of K-selected equilibrium species, although r-selected species, such as bacteria with their relatively short genetic programs and high turnover rates, were also able to cope with change.

Behavioral Versatility and Survival

Biologist Daniel Sol and colleagues have evidence that large brain size confers a behavioral flexibility that can have long-term survival benefits. A study of 66 species of mammals that were introduced into novel environments on both islands and mainland found higher survival success for species with larger relative brain size.[12] *Relative brain size*, here and elsewhere, means having a larger brain than an average animal of that particular body size. Among birds in their native habitats, larger-brained species also have higher annual survival rates in all kinds of environments, from the poles to the tropics.[13] Birds that stayed as year-round residents rather than migrating have innovative ways of feeding in winter, and this versatility also correlated with larger relative brain size.[14] Sol suggests that large-brained species are better at tracking resource variation, dealing with environmental complexity and change, and avoiding unfamiliar predators.[15]

Overall, a large brain affords behavioral flexibility that allows a species to deal with changes in its surroundings, a theory that Sol calls the *cognitive buffer hypothesis*.[16] The brain size of modern birds and mammals is about tenfold the brain size of vertebrate groups that evolved before them.[17] In the great extinction that marks the boundary between the Mesozoic and Cenozoic eras, both groups showed high survival rates. Recent data show that all the living orders of placental mammals were present 75 mya, a good 10 million years before the asteroid struck, and that they survived the catastrophe, so that the "mass extinction event had, at best, a minor role in driving the diversification of the present-day mammalian lineages."[18] Birds date back at least as far as Archaeopteryx 150 mya, and recent DNA analysis indicates that at least 24 bird families originated before the asteroid impact—and 15 of these families are still with us.[19] There is a gap between species numbers predicted by molecular studies and what has been found in the fossil record for birds, but it is clear that a variety of modern bird orders was already present 10 million years before the great extinction and that they persisted afterward.[20]

The end-Mesozoic extinction wiped out 100 percent of the dinosaurs, but only 6 percent of lizard and snake families perished, so being small and widely distributed were also likely factors in survival. Only 14 percent of placental mammal families were lost, but mammals at the time generally were also small-bodied.[21] Most mammals and birds had not yet reached the brain size of modern species, although they had increased beyond that of reptiles of similar body size long before the extinction event.[22] The early mammals of the Mesozoic already had an average twofold increase in relative brain size over their nearest non-mammalian ancestors, the cynodonts.[23] All mammalian brains have a six-layered neocortex that is the seat of most learning and executive functions. Reptiles have a similar area known as the *dorsal cortex*, but it is only two cell layers thick and probably does not have as much processing power.[24] Early bird species also likely had brains significantly larger than those of reptiles of similar body weight.[25] This may have conferred a behavioral versatility that allowed new ways of feeding and finding shelter in the chaotic conditions following the asteroid impact, and being warm-blooded would have buffered them from the variable climate that followed. Increased homeostasis, therefore, may have long-term survival value.

Primate Evolution and the Environment

A K-selected strategy of increased behavioral versatility and increased cognitive powers may have been essential in the lineage that led to us. Paleoanthropologist Richard Potts of the Smithsonian Institution proposes that complex environmental change was the cause of cognitive increases of primates in general and especially in the hominid lineage. He notes that during the Miocene epoch (23 to 5 mya), there were large seasonal climate shifts that led to the fragmentation of forests and differing periods of ripening of the fruits on which the great apes depended. During this period, primate teeth show greater dental and dietary diversity. They apparently responded to the increased unpredictability of the environment with greater behavioral flexibility and increased cognitive abilities. Complex mental representation would have allowed them to keep track of widely scattered food resources that ripened at different times, and fluid group dynamics would have led to different-size social groups with the best chances for survival.[26] Potts states: "Thus lineages build up adaptability to environmental change, manifested in cognitive, social, and anatomical characteristics that enable organisms to absorb and process information about their surroundings, and to respond in ways that enhance their survival and genetic contribution."[27]

The hominid lineage leading to our own species carried these trends further. Climate fluctuations grew more extreme from the Miocene to the Recent, and hominid evolution coincided with the largest climatic oscillation of the late Cenozoic.[28] Key adaptations such as early bipedalism, encephalized brains, and complex social groupings allowed hominids to exploit new food resources and explore wider territories. During the early Pleistocene epoch (2 mya) there is evidence of more varied tool making in a wider range of habitats, with rocks being transported over greater distances. The greatest increase in brain size began with the onset of the largest climate oscillation of the past 6 million years. According to Potts, "Relative expansion of the brain enhanced its data processing and integrating functions, which greatly improved versatility and responsiveness to novel adaptive problems."[29] Potts proposes that increased mental abilities to represent the environment also led to greater abilities to understand the mind of others (the "theory of mind") and higher levels of self-conception.[30]

Plant Variability

The Cenozoic era is known as the *age of mammals* because placental species radiated widely during that period. But it is also the age of birds and flowering plants (angiosperms), because both these groups diversified greatly in the new era. Like the mammals, they originated much earlier in the Mesozoic and were already diversifying with many new species when the asteroid struck. That event led to the loss of 57 percent of plant species in North America and 18 to 30 percent at the family level. But few or no flowering plant families were lost.[31] As with mammals and birds, the disaster merely punctuated a diversification that was already under way. There is considerable debate over the reasons for angiosperm success, but I would like to suggest that their survival may have been due to similar reasons as that of the mammals and birds; that is, they have a behavioral versatility that buffers them from fluctuations in their environment.

In the first place, angiosperms have a developmental plasticity that allows them to colonize many different niches. For example, the Hawaiian silverswords are related to sunflowers on the mainland. From a common ancestor that arrived on the islands about 5 mya, they have radiated into 28 species that take every form from small plants to shrubs, trees, and vines that live in nearly every habitat across the islands.[32] Overall, angiosperms constitute nearly 90 percent of all plant species on land today, span every body plan from tiny aquatic duckweeds to large trees, and occupy every habitat on Earth. Their main former competitors from the Mesozoic, the gymnosperms, have been reduced to 0.29 percent of land plant species, though they still comprise some large northern forests. The earlier-evolving ferns now make up 3.99 percent of plant species.[33]

Besides developmental plasticity, angiosperms can borrow from the nervous systems of others. By putting out colorful flowers and nutritious nectar, they lure a variety of animals to aid in pollination and thereby overcome the immobility of plants. Insects tend to stay species-specific when they visit flowers, and this keeps pollination efficient. Gymnosperms, which rely on the wind to carry pollen to female ovules, have to grow in large clumps to keep fertilization efficient, but angiosperms can be more widely scattered. The number of flowering plant species and the number of insect families are highly correlated in the fossil record, showing their symbiotic relationship.[34]

Fruit is also a unique angiosperm invention that allows them to enlist a variety of animals in distributing their seeds. Fruit, like nectar, comes cheaply to plants because sugar is the primary product of photosynthesis. Typically, hard indigestible seeds are buried inside fleshy fruit. These pass unscathed through the animal digestive system and are deposited far from the parent plant inside animal droppings that make good fertilizer. Some flowers have even enlisted bacteria to help in their life cycle. Nine families of angiosperms attract *rhizobia* to their roots, tiny soil-living bacteria that can fix nitrogen and supply it to the growing plants, which reciprocate by feeding carbohydrates to the bacteria inside special chambers known as *root nodules*. Because this process works better in a low-oxygen environment, these plants have evolved a special molecule to carry away oxygen from the root nodules called *leghemoglobin*. Some nodules look red because this molecule is similar to the hemoglobin we use in our red blood cells.[35] No gymnosperms have been able to evolve these nitrogen-fixing alliances.[36]

Without having to evolve a nervous system, flowers have been able to make use of the nervous system of animals and thereby broaden their opportunities to reproduce and disperse. The plant world has become more personalized, in the relation both of one plant to another and of plants to the animal world, and these alliances account for a large part of angiosperm success. The versatility of this group goes further. Some have returned to simpler ways of life, like the grasses that dispense with nectar and rely on wind pollination. Many flowers have evolved an annual habit and are able to grow and reproduce within one year. This r-selected strategy allows them to colonize new habitats quickly and outcompete the slower-growing gymnosperms and ferns.

Since flowers manage behavioral versatility without nervous systems, they must rely almost entirely on gene expression for fundamental changes in behavior. An animal can run away from a predator; a plant makes secondary compounds like poisons or hallucinogens to deter herbivores. An animal can walk to food; a plant must grow toward it. Plants can use genetic mechanisms instead of memories that would be stored in a nervous system. For example, many plants have to go through a period of cold to be able to flower in the spring. Recent work shows that plants release molecules in response to prolonged cold that alters how their DNA is folded, allowing the genes for flowering to be expressed.[37] The reliance on gene expression to make behavioral changes may be the reason why we find higher gene counts in plants than we do in animals.

Plants appear to be much more variable than animals in their use of genes. Angiosperms vary 2,000-fold in the size of their genomes, whereas mammals only vary fivefold. Flowers can have up to 640 chromosomes per cell, whereas mammals have a maximum of 134.[38] Angiosperms have also increased their diversification by whole genome duplication, a condition known as *polyploidy*. When mammals make babies, the egg and sperm each carry half of the parent's chromosomes. Gametes go through reduction divisions before they are released, a process known as *meiosis*. In plants, through errors in meiosis, unreduced gametes can be produced. When they unite, the embryo will have twice the normal number of chromosomes. In most animals, this would be fatal for the embryo. But plants can survive and in fact build taller, stronger plants by way of polyploidy. It is estimated that 70 percent of angiosperms have gone through at least one round of polyploidy in their ancestry.[39]

Some research suggests that angiosperms control nutrient levels in their habitats by the decomposition of their plant bodies and the leaves they drop.[40] The high growth rates of flowering plants require high nutrient levels in the soil. Gymnosperms do better in soils with low nutrients. Angiosperms keep higher concentrations of both nitrogen and phosphorus in their leaves and return these to the soil with the deciduous habit of dropping their leaves (another angiosperm innovation). There may be a tipping point at which angiosperms become numerous enough to maintain soil conditions that are optimal for their own growth.[41] A global survey of flowering plants (eudicots) found that their litter decomposed faster than that of plant groups that evolved before them: an average four times faster than the litter of mosses (bryophytes), three times faster than that of ferns, and 1.8 times faster than that of gymnosperms.[42] Angiosperms are thereby increasing the cycling of nutrients and energy with their high growth rates and quick replacement, and they are maintaining an environment that is advantageous to their own growth, a kind of homeostatic strategy.

During much of the Mesozoic, the world was warm and moist, and most areas enjoyed a tropical environment. Pangea was at its most extensive, there was no ice at the upper latitudes, and dinosaurs were able to spread from the equator to the poles.[43] During the later half of the Mesozoic, the continents began to break up, and by the end of the Mesozoic they were approaching their current positions (see figure 2.1). This fragmentation increased the variety of habitats found on different continents, and this increase favored the fast-speciating angiosperms.[44] Climate was also cooling in the late Mesozoic, and this process increased in the Cenozoic. About 34 mya, strong seasonality returned with shorter and more severe fluctuations between warm and cold.[45] In this era, which extends to the present, the most homeostatic groups are dominant—those that can buffer themselves against fluctuations in their physical surroundings. Endothermic animals do this by regulating body temperature, and both flowering plants and animals engage in increased nurturing and protection of their young. But perhaps the greatest homeostatic strategy is behavior— that is, being able to innovate and change over short periods to meet the challenges of a fluctuating environment.

The Pace of Life

There has been a speedup in the pace of life over the long range of evolutionary time. Bacteria (prokaryotes) were the first and only life forms on the planet for almost 2 billion years. Bacteria can live at 1 percent of current oxygen levels, but eukaryotes (cells with nuclei) need at least 2 to 3 percent of current oxygen concentrations to live.[46] Using oxygen, a cell can get about 16 times more energy from a molecule of fuel (glucose) than it can without oxygen. Without oxygen, a cell can metabolize a molecule of glucose to produce only two of adenosine triphosphate (ATP, the main energy currency of cells). But with oxygen it can break the leftovers down to yield 30 more ATP molecules. Oxygen allows for high-octane living.

But free oxygen was a scarce molecule in the early days of planet Earth, and its concentration in our atmosphere built up slowly. Oxygen is highly reactive, and in the heat of planet formation most of it was bound up in rocks or in the production of water. Only after the first photosynthetic bacteria evolved about 2.5 billion years ago was molecular oxygen slowly released. A worldwide band of reddish iron deposits testifies to the slow oxidation of iron in the early oceans by this liberated oxygen. After this, its concentration began to increase in the atmosphere to the 21 percent we have today.

Large eukaryotic cells evolved when oxygen concentration was at about 3 percent of current levels, and multicellular eukaryotic organisms

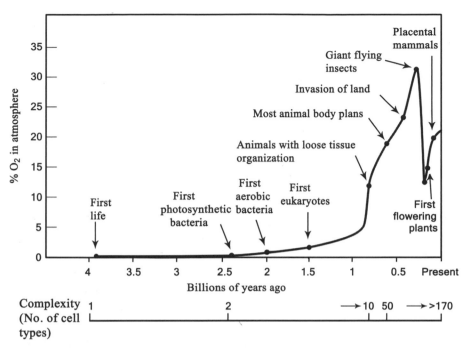

Figure 2.2 Oxygen concentrations and the history of life.
Rising oxygen levels support increasing complexity of structure, as measured by the number of different cell types in an organism. (Data on cell types: John Tyler Bonner, *The Evolution of Complexity by Means of Natural Selection* [Princeton: Princeton University Press, 1988]; J. W. Valentine, A. G. Collins, and C. P. Meyer, "Morphological Complexity Increase in Metazoans," *Paleobiology* 20, no. 2 [1994]: 131–142; V. J. Thannickal, "Oxygen in the Evolution of Complex Life and the Price We Pay," *American Journal of Respiratory Cell and Molecular Biology* 40, no. 5 [2009]: 507–510. Data on oxygen levels: H. D. Holland, "The Oxygenation of the Atmosphere and Oceans," *Philosophical Transactions of the Royal Society B—Biological Sciences* 361, no. 1470 [2006]: 903–915; David E. Sadava et al., *Life, the Science of Biology*, 9th ed. [Sunderland, Mass.: Sinauer, 2011]. Data on dates: R. A. Berner, J. M. Vandenbrooks, and P. D. Ward, "Evolution: Oxygen and Evolution," *Science* 316, no. 5824 [2007]: 557–558; E. H. Davidson and D. H. Erwin, "An Integrated View of Precambrian Eumetazoan Evolution," Cold Spring Harbor Symposia on Quantitative Biology 74 [2009]: 65–80)

appeared at about 5 percent current oxygen levels. Figure 2.2 shows important events in the history of life in relation to oxygen concentrations.

Nearly all eukaryotes need oxygen to survive, whereas many prokaryotes (bacteria) can live without it. Large cells (and bodies made of multiple cells) have less surface area per unit volume than single small cells do, and that in itself requires higher oxygen levels to meet metabolic needs. But eukaryotic cells are also more complex than prokaryotes in nearly every way, with more DNA, more compartments, and more cell organelles. Their higher complexity probably requires the higher energy that metabolism with oxygen (aerobic respiration) makes possible. Biologists Nick Lane and William Martin calculate that a eukaryotic cell has about 1,900 times more power available per gene than an average prokaryotic cell.[47] This is due to the presence of power stations inside eukaryotic cells, known as *mitochondria*. These organelles are the descendents of an ancient prokaryote that took up residence in the eukaryotic ancestor and, by using oxygen, could take the remnants of metabolism and produce extra ATP for its host. This was perhaps the defining quality that allowed eukaryotes to go on to greater complexity of organization. Eukaryotes invented some 3,000 novel gene families and five times as many protein folds as their prokaryotic predecessors. Although replicating DNA costs only 2 percent of a cell's energy budget, protein synthesis uses 75 percent. With their vastly expanded energy reserves, eukaryotes could undertake much more complex protein architecture. At least six eukaryotic lineages made the transition to multicellularity, whereas prokaryotes never got beyond rudimentary levels of association.[48]

Key events in the increasing complexity of life correlate with bursts of increasing oxygen levels. During the Cambrian explosion, beginning about 535 mya, most modern body plans of animals appeared in a fairly short time, along with many forms that quickly went extinct. Shortly before this time there was a rapid increase of oxygen levels that saturated deep into the oceans. Researchers suggest this may have been the first time oxygen levels were sufficient to support the metabolism of large multicellular animals.[49] Rapid increases of oxygen are also associated with the conquest of the land by insects at 410 mya and with the evolution of large-bodied mammals.[50]

Beginning in the Carboniferous period (359 mya), extensive swamp forests of giant tree ferns and horsetails grew. Their accumulated debris formed the coal deposits we find today. Burial of this organic matter protected it from oxidation, so free oxygen increased in the atmosphere. Together with oxygen that is the normal byproduct of photosynthesis, this led to levels of atmospheric oxygen not seen before or since. Giant insects and amphibians flourished. These populations crashed at the end of the Permian period (251 mya), as the continents came together to form the supercontinent Pangea. Swamps dried out, so fewer plants grew or were buried. Oxygen levels quickly dropped from about 30 percent to 15 percent. Together with massive volcanic eruptions and general cooling of the climate, this led to the greatest extinction of animal and plant life the planet has ever seen.[51]

Oxygen levels gradually rebounded, and modern placental mammals diversified between 100 and 65 mya, a period during which oxygen increased about 50 percent from its lows.[52] By way of the placenta, a mother's oxygen is transferred to the fetus's bloodstream, and high atmospheric oxygen is necessary to make this exchange efficient. Most modern mammals cannot reproduce above roughly 4.5 kilometers' (2.8 miles') elevation, which corresponds to oxygen levels in the period before placental mammals evolved.

The rise in oxygen is related to increases in the complexity of body plans. This is represented by the number of cell types found in organisms, as shown at the bottom of figure 2.2. Life remained essentially single celled until oxygen reached about 5 percent concentration in the atmosphere (a cell count of two refers to an active cell and its dormant form as a spore). As already noted, more complex eukaryotic cells require higher ambient oxygen levels than bacteria do. Animals with loose tissue organization, such as sponges and jellyfish, evolved at oxygen levels of about 5 percent and are estimated to have about 10 different cell types. At the Cambrian explosion, when most animal body plans appeared, oxygen had risen to about 18 percent, and about 50 cell types are estimated for the most complex organisms. After the end-Permian crash of oxygen, placental mammals diversified at about 20 percent oxygen, and vertebrate bodies are estimated to have about 170 cell types (not counting different kinds of neurons).

Note the speedup of time in figure 2.2. It took some 2 billion years to get from single-celled prokaryotes to single-celled eukaryotes. Multicellular life became widespread only in about the most recent 15 percent of the history of life on our planet, and there has been an accelerating pace in the complexity of body plans as measured by the variety of cell types found in an organism.

The relationship between complexity and increasing energy use applies to both the organic and inorganic realms. Chapter 9, "Energy Flows," addresses the theories of Eric Chaisson, who shows how inanimate forms like stars become more complex in structure as energy passes through them at greater rates. He finds common principles of increasing complexity and energy use in living and nonliving forms.

Circulation Changes

An increasing pace of life can be found in the locomotion of animals. The first land vertebrates probably moved slowly with side-to-side undulations, as we still see in crocodiles and many salamanders. This is a relic of the undulatory motions that fish use to swim. In the lineages leading to dinosaurs, birds, and mammals, legs moved more vertically under the body, allowing faster motion. Standing upright decoupled locomotion and breathing and allowed expanded use of oxygen for metabolism. Birds and mammals evolved this independently.[53]

Circulation also changed to allow higher activity levels. Fish have a two-chambered heart, and oxygenated blood from the gills goes directly to the body (the systemic circuit). Amphibians and reptiles have a three-chambered heart, allowing the heart to pump blood separately to the lungs and the body. However, some mixing of oxygen-rich and oxygen-poor blood can take place in the heart. Birds and mammals have, in effect, a double heart (figure 2.3). One side is entirely devoted to pumping blood to the lungs, and the other side pumps only to the body, so that high pressure is maintained in both systems and oxygen-rich blood is not diluted with oxygen-poor blood. This circulation serves the constant high metabolic needs of birds and mammals.

Mammals and birds have about a 10-fold increase in brain size over that of vertebrate groups that evolved before them. The brain is an expensive organ. In humans, it occupies only 2 percent of the body but accounts for about 20 percent of our basal metabolism. It has eight to 10 times the metabolic cost of skeletal muscle per unit mass.[54] A recent study of 1,247 mammal species found a positive correlation between relative brain size and basal metabolism rate.[55] An increase in basal metabolism also means that it becomes more expensive to live: an animal must burn more calories just to maintain itself. Reptiles and amphibians in general need to feed less often than birds and mammals, both of which must constantly forage for food.

The greater expense of living must come with some advantage. Perhaps large brain size gives dominant animals an advantage in an increasingly complex environment. The diversity of life has increased over the long reach of evolutionary time. Through the Mesozoic (the age of dinosaurs) and especially in the Cenozoic (the age of mammals), the pace of diversification has quickened in both land plants and land animals, de-

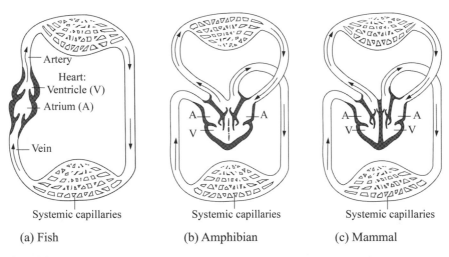

Figure 2.3 Circulatory systems.
(*a*) A fish heart has one atrium and one ventricle. (*b*) Amphibians have a three-chambered heart with a partial division in the ventricle that allows some mixing of oxygen-rich and oxygen-poor blood. (*c*) In mammals and birds, a four-chambered heart completely separates circulation to the lungs (*top*) from circulation to the rest of the body (the systemic capillaries). (From Neil A. Campbell and Jane B. Reece, *Biology*, 6th ed. [San Francisco: Benjamin Cummings, 2002], 874. Adapted by permission of the publisher)

spite periods of extinction, leading to a nearly threefold rise in the number of families in both groups.[56] Perhaps the accelerating pace of life is related to the growth of information content both within organisms and in an increasingly diverse environment of animals and plants. A more complex environment may have required more complex mapping in brains to cope with it. In this sense, information may beget information, at an accelerating rate, like the accumulation of compound interest. The complexity of body plans has grown more and more quickly, its pace measured by the number of different cell types. This growth has been accompanied by an increase in the number of genes, especially those guiding body architecture, as detailed in the next chapter. There has also been an expansion of relative brain size in mammals and birds over that of groups that evolved before them. The increased complexity of the external environment may be related to the increased complexity of gene networks, and brains needed to respond to that environment.

The Modern World

Many of the concepts we have considered so far can be seen at work in the modern world. The parallels may be simply fortuitous, or they may say something about the gathering complexity of organic systems in general. Industrialized countries are more information intensive and more homeostatic than underdeveloped countries. The pursuit of knowledge has led to improved ways of fighting disease, made food plentiful, and allowed swift communication and travel. In our large cities we heat and cool buildings at will, the lights never go out, and the traffic never stops. All of this takes a lot of energy and leads to an ever-accelerating pace of life. It is clear here that information begets information. Each field has to adjust as new discoveries appear in related fields. The number of technical journals is continually multiplying, so that no one person can keep up anymore, and increasingly what an individual has to sell is his or her expertise in some specialized field. For similar reasons, the pace of competition is continuously increasing in business. The industrialized world bears the marks of K-selection: we live longer and put off reproduction, in part because it takes so long to train the young in the skills of society. Education becomes so expensive that a family simply cannot afford to have many children.

The developing world is more r-selected and less homeostatic. People tend to have larger families and shorter life spans. Parents have large families in part because they will lose a certain number of children to infant mortality and they expect the extra hands to help with chores around the house and farm. Life is more subject to the vagaries of weather, famine, and disease. It moves at a slower pace and demands less of concentrated energy sources.

In succeeding chapters, I will explore further the connections between these concepts. Homeostasis is a K-selected strategy. It allows a species to maintain steady population numbers in a fluctuating environment, and behavioral versatility is a key strategy in dealing with change. Mammals and birds have high caloric needs and are also the most homeostatic in terms of the varied conditions in which they can operate. They have high information content in terms of genes and relative brain size, concepts that are defined more precisely in the next chapter.

3

Information Content

Life gathers information in genes and brains, and in both there has been an increase in information content over time. There has been an increase in the average number of genes in going from prokaryotic cells (without nuclei) to eukaryotes (with nuclei), and a further increase in going from multicellular organisms with loose tissue organization to animals and plants with complex tissues. There has also been an increase in the size of nervous systems from their first appearance in the diffuse nerve net of jellyfish to the centralized control of insects and vertebrates. This is measured in terms of the weight of the brain compared to the weight of the body it controls. In both information-gathering systems—genes and the neurons of nervous systems—microprocessors have developed mechanisms for integrating information from many sources to make a decision.

Gene Counts

Prokaryotic cells, which held sway on the planet for about the first 2 billion years of life, generally have anywhere from about 350 to 8,000 genes, whereas animals and plants contain more than 13,000 genes.[1] The difference in gene count between prokaryotes and eukaryotes is related clearly to structural complexity. Eukaryotic cells are larger and more complex in almost every way: more membranes, more compartments and stages in building proteins, more complex skeletal support (cytoskeleton), more cell surface molecules, and more complex ways of controlling genes. Some of a eukaryote's largest subunits (organelles) are derived from prokaryotic cells that took up symbiotic residence inside the eukaryotic ancestor: the mito-chondrion, which is the main source of energy, and the chloroplast, which is the site of photosynthesis.

The transition from prokaryotes to eukaryotes is the clearest example of the connection between information content and complexity of struc-ture. The higher gene counts of eukaryotes allowed more internal orga-nization along with an expanded protein repertoire of some 3,000 novel gene families, with proteins that are on average about 30 percent longer and have about five times as many new ways to fold.[2] In some ways, the transition from prokaryotes to single-celled eukaryotes required more innovations than did multicellular body plans that came later. About 40 percent of protein families in eukaryotes are novel compared with those in prokaryotes, whereas only about 20 percent of protein families in multicellular eukaryotes are new compared with those found in single-celled eukaryotes.[3]

The transition to eukaryotes also shows a relationship between com-plexity and energy use. Eukaryotic cells have about 1,900 times more power available per gene than an average prokaryotic cell.[4] Issues of complexity and energy use will be explored further in the discussion of thermodynamics in chapter 9.

A study of 682 prokaryotic genomes found that as the total gene count goes up, the number of genes for a special class of proteins that regulate gene expression, known as *transcription factors*, increases much faster than that of genes as a whole. They increase with an exponent of 2, meaning that as the total gene count doubles, gene regulatory proteins increase by a factor of 4.[5] A similar study of a smaller number of eukaryotic genomes found similar results.[6] This indicates that increases in gene regulation may be key to understanding increases in overall organismal complexity.

Although the first drafts for a variety of eukaryotic species are now available, they come with considerable uncertainty about total gene counts. There are a variety of methods for finding genes, and each has its own limitations. Some rely on signals for the start and editing of genes, others on actual proteins produced in the cell, and others on gene comparisons between species. The different methods often do not agree in their tal-lies. The first complete draft of the human genome, for example, came out in 2003, yet in 2010 there is still debate over some 2,000 genes.[7] The data presented in table 3.1 come mainly from Ensembl,[8] which is a joint project of the European Bioinformatics Institute and the Wellcome Trust Sanger Institute. Their numbers combine information from a variety of

Table 3.1. Gene Counts for a Variety of Taxa and Their MicroRNA Totals

Common Name	Scientific Name	Protein-Coding Genes	MicroRNAs
Prokaryotes			
Spherical bacterium	*Staphylococcus aureus*	2,660	
Rod-shaped bacterium	*Bacillus subtilis*	4,176	
Intestinal bacterium	*Escherichia coli* K12	4,262	
Eukaryotes	**Single-celled**		
Baker's yeast	*Saccaromyces cerevisiae*	6,696	0
Brown alga	*Phaeodactylum tricornutum*	10,402	
Diatom	*Thalassiosira pseudonana*	11,674	
Green alga*	*Chlamydomonas reinhardtii*	14,516	50
Eukaryotes	**Loose tissue organization**		
Social amoeba	*Dictyostelium discoideum*	17,635	2
Bread mold	*Neurospora crassa*	9,820	0
Soil fungus	*Aspergillus nidulans*	10,534	
Invertebrates	**Complex tissues**		
Fruit fly	*Drosophila melanogaster*	14,076	171
Mosquito	*Anopheles gambiae*	12,604	67
Roundworm	*Caenorhabditis elegans*	20,158	175
Sea squirt	*Ciona intestinalis*	14,180	332
Vertebrates			
Human	*Homo sapiens*	22,320	940
Chimpanzee**	*Pan troglodytes*	19,829	606
Mouse	*Mus musculus*	23,062	590
Chicken*	*Gallus gallus*	16,736	499
Zebrafish*	*Danio rerio*	24,147	360
Plants			
Flower	*Arabidopsis thaliana*	31,280	199
Cereal	*Sorghum bicolor*	34,496	148
Rice	*Oryza sativa*	57,995	447
Poplar tree	*Populus trichocarpa*	41,377	234

NOTE: For species with either a small (*) or large (**) percentage of projected genes as of January 2011, totals cannot yet be accepted with certainty.
SOURCE: Gene data from Ensembl: Assembly and Genebuild, http://useast.ensembl.org/index .html; plant data from EnsemblPlants, http://plants.ensembl.org/index.html; microRNA totals from http://www.mirbase.org/cgi-bin/browse.pl (all accessed January 10, 2011).

sequencing methods to arrive at a total gene count. Species that are starred have either a small (*) or large (**) percentage of projected genes as of January 2011, so their totals cannot yet be accepted with certainty. Even the best counts may still vary by 5 to 10 percent in the future.

The results in table 3.1 show an increase in gene counts in going from prokaryotes to eukaryotes. There is a further increase in going from multicellular eukaryotes with loose tissue organization to those with definite organ systems. *Loose tissue organization* refers to eukaryotes that live in colony-like arrangements but do not build true tissues, such as many fungi. But there are several surprising results in these totals. Vertebrates have only about twice as many genes as insects have, and it takes about the same number of genes to build any vertebrate body, from fish to human. Moreover, plants regularly have more genes than animals, and in the case of rice, more than twice as many.

It may be that gene counts are a poor way to predict organismal complexity. But we may be thinking too simply on these issues. There are increases in the number of regulatory genes that correlate with greater body complexity. A tool kit for manipulating genes appears early in evolution, but it is used to build increasingly complex structures. Because regulatory genes make up a small percentage of total genes, we can have growth in complexity that does not affect the total gene count by much. The microribonucleic acids (microRNAs) listed in table 3.1 are also part of the regulatory system of genes that will be explained later in the chapter.

Also, I suggest that we need to consider another way in which life gathers information: in nervous systems. These are relatively open-ended mechanisms that can increase in size without increasing in gene number. For example, humans and mice have about the same number of genes, but the relative brain size of humans has risen about ninefold. Plants do not have nervous systems; they mainly make changes in behavior through changes in gene expression. So much behavior must be encoded in the genome of plants, and this may explain why they have higher gene counts than animals.

Gene Control

To understand changes in the complexity of gene control, we need to understand how genes are regulated. Every gene comes with a gene switch, just as every light comes with a light switch. It would be wasteful, and downright harmful, to run all your genes all the time, just as it would be wasteful to burn the lights in every room of the house all the time. The control region of a gene generally is upstream of the structural part of a gene (the sequence that codes for the protein it will produce), and it may cover thousands of base pairs of DNA. The region nearest the structural part of the gene is called the *promoter*. This is the place where RNA polymerase lands, begins prying open the double-stranded DNA, and begins to synthesize (polymerize) a strand of RNA that is complementary to the code it reads on one side of the DNA (figure 3.1). This RNA copy will then be used to produce a protein. RNA polymerase will continue through the structural gene until it reaches a stop signal that causes it to fall off the DNA. The main molecule that produces RNA for proteins in eukaryotes is known as *RNA polymerase II*.

It turns out that RNA polymerase, by itself, has poor affinity for DNA. To locate a specific promoter for a particular gene, it needs helpers. These helpers go by the general term of *transcription factors*, and in eukaryotes there can be a lot of them. Together they form an assemblage that pulls RNA polymerase out of solution, positions it at the head of a promoter, and then releases it to begin its journey down the DNA to produce an RNA message. This assemblage is the main point of control for genes, the main switch that decides whether a gene is active or not, and its complexity allows a sophisticated level of control.

Part of this assemblage may originate far away from the promoter. As shown in figure 3.1, activator proteins may bind far upstream from the promoter. They recognize specific sequences in the DNA, called *enhancer sequences*, and, in the current model, bend over to interact with the other transcription factors at the promoter. The entire assemblage acts to grab RNA polymerase and position it at the start of the gene. This assemblage can have various levels of efficiency. As shown at the top of figure 3.1, a repressor protein might compete with activator 3 by recognizing a sequence that overlaps with enhancer 3. If the repressor gets there first or is in higher concentration than activator 3, its effect may be to repress the entire assemblage and prevent it from efficiently recruiting RNA polymerase.

The transcription apparatus, which may comprise a hundred subunits or more, is in fact a microprocessor, combining enhancer and repressor subunits and building a structure that is more or less efficient at grabbing RNA polymerase and positioning it at the start of a gene. Any missing subunit, or the inclusion of one or more repressors, could be a

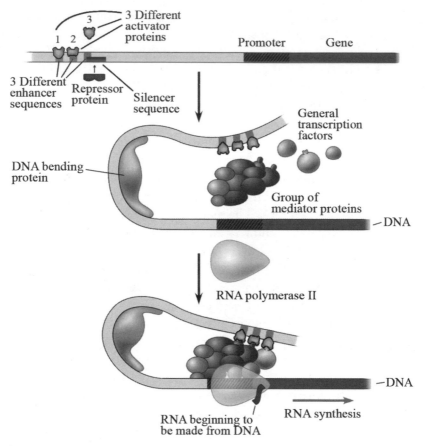

Figure 3.1 Gene control in eukaryotes.
Multiple factors must come together to build an apparatus that positions RNA polymerase at the start of a gene. A repressor protein or a missing factor can make the apparatus inefficient, producing little or no RNA output from the gene. (From Neil A. Campbell et al., *Biology*, 8th ed. [San Francisco: Pearson Benjamin Cummings, 2008], 360. Adapted by permission of the publisher)

deal breaker and make the overall apparatus nonfunctional. On the other hand, just the right set of activators might make a gene highly active and very efficient at binding one RNA polymerase after another and sending it down the gene to produce many RNA messages.

These activating and repressing proteins may come from neighboring cells, and they enable the complex patterns we find in multicellular bodies. Think, for example, of how your hand transitions from palm to finger to nail. What allows cells in different areas to produce different proteins? The answer, in large part, depends on transcription factors they send to each other. At the nail, activator proteins have turned on the genes for nails but turned off muscle genes, whereas the opposite is taking place in muscle cells of the finger itself. Neighbor cells can send to each other messages that add up to different accumulations of transcription factors in different parts of the body and that turn on different sets of genes. Studies show that these mechanisms can be precise to a width of only one cell. That is, if row A of cells is sending a repressor protein to row B, row A can be in a different set of gene activities than row B. These mechanisms are especially important in early development, when embryonic cells are making decisions to become different parts of the body, such as muscle, nerves, or bone. They begin with the potential to be many things, but like trains leaving a common station house, they set out on different tracks, and the switching mechanisms that decide one track or another are different groups of transcription factors.

Prokaryotic cells, with their simpler cell structure, also have much less sophisticated control of their genes than eukaryotes do. There are fewer repressors and enhancers to interact with RNA polymerase at each particular gene, and fewer transcription factors overall. The common bacterium *Escherichia coli* has a total of about 300 such factors, whereas humans have some 2,000.[9] Also, whereas eukaryotes control one gene with one promoter, prokaryotes bunch genes that have a common function together under the control of a single promoter. For example, in bacteria, three genes needed to digest the sugar lactose are grouped together under the control of a single promoter in a structure called the *lac operon*. Operons are common in prokaryotes for a variety of functions, whereas the general rule in eukaryotes is one promoter to one gene.

The Evolution of Regulation

Changes in the complexity of gene regulation appear to be key to understanding changes in body complexity in evolution. A study looked at 1,219 protein families in 38 eukaryotic species that varied in complexity according to the number of different cell types that could be identified in their tissues. The study's scope ranged from single-celled protists through fungi, insects, and vertebrates, including humans. Protein families were classified according to functional domains in their three-dimensional structure.[10] Only 15 percent of these protein families showed correlation with increases in body complexity, but the ones that did expand were instructive. Fifty percent of them were involved either in gene regulation or cell communication. Cell communication includes signals between cells, within cells, and adhesion molecules between cells. The other 85 percent of gene families did not show changes that correlated with increases in body complexity. This seems to indicate that it is not the basic building materials that change with more complex architecture but the ways in which they are put together. The complexity of instructions for putting the basic materials together does increase, along with ways in which subunits relate to each other. In all, about 200 protein families grew in abundance with increased body complexity, as measured by the number of different cell types in that body.[11]

Other studies show similar results. Estimates of the size of regulatory regions next to individual genes show increases from organisms with loose tissue organization such as fungi to invertebrates to mammals.[12] Also, the proportion of genes that encode transcription factors rises from about 5 percent in fruit flies and nematode worms to almost 10 percent in mice and humans.[13] An expansion of transcription factor classes happened early in animal evolution. Ten classes of transcription factors have been found in sponges, near the base of the animal family tree. This compares with 88 classes imputed to the last common ancestor of later, bilateral animals (in bilateral animals the left side is a mirror image of the right), and about 108 classes for the last common ancestor shared by insects and vertebrates (the protostome-deuterostome last common ancestor).[14]

Early Body Forms

The transition from single-celled eukaryotes to more colonial associations did not in itself appear to require much in terms of extra genes. *Chlamydomonas* is a single-celled green alga that is able to move about using two whiplike flagella on one end. *Volvox* is one of a group of multicellular green algae composed of associations of cells very similar to *Chlamydomonas*. Species vary from simple colonies of a few cells stuck together in a jellylike matrix to large spheres of thousands of *Chlamydomonas*-like cells that tumble through the water by beating their flagella. *Volvox carteri* has about 2,000 cells with the simplest of distinctions: the outer ones of the sphere do the rowing while 16 large cells toward one end of the interior reproduce. This primitive multicellular organism therefore makes a fundamental distinction found in animal life. One group of (somatic) cells builds a body while a select group of (reproductive) cells is devoted to perpetuating the species. The reproductive cells of *V. carteri* multiply to build little spheres inside the parent sphere, which eventually opens, releases the young, and then dies.

The single-celled *Chlamydomonas reinhardtii* has been sequenced at 14,516 genes,[15] and the first draft of *Volvox carteri* comes in at 14,520 genes, with an overlap in most of their genes families. About 12 percent (1,835) of the protein families do show striking differences between the two species, with more members in *Volvox*. But it does not appear to take more genes in absolute numbers to go from single-celled life to simple colonial forms, at least among the green algae.[16]

The body plan of sponges has changed little since the dawn of the Cambrian period, about 540 million years ago (mya). Sponges lack true organ systems, muscles, or nerves. Their bodies are so loosely organized that their cells can be dispersed with a sieve and the body will reaggregate on the other side. Sponges are filter feeders that draw water into their cuplike bodies and strain out nutrients suspended in the water. They create water currents with cells that line the interior body cavity. These cells, called *choanocytes*, extend a whiplike flagellum whose undulating motion moves water through the porous walls of the sponge.

The choanocytes are similar to a group of free-living single-celled eukaryotes known as *choanoflagellates*. In terms of genetics and morphology, this group is now considered the closest known relative of multicellular animals. One of these single-celled organisms, *Monosiga brevicollis*, has now been sequenced, and it is estimated to have about 9,200 genes.[17] It produces a repertoire of transcription factors that is more restricted than that of multicellular animals. It has five of ten transcription factor families found in animals and four of six cell adhesion domains. It does not contain any of the Hox genes used by all animals to define segments along a body plan, although it does have genes with homeodomains, an ancient protein motif that can recognize sequences on DNA. The choanoflagellate also lacks Toll genes that are an important signaling system in the embryonic development of animals. Surprisingly, this single-celled protist does have genes for a variety of cell adhesion molecules previously thought restricted to multicellular animals, and it probably uses them to attach to its external environment.[18]

The multicellular sponge *Amphimedon queenslandica* has been sequenced, and the first draft estimates about 18,700 protein coding genes.[19] Despite its simple body plan, it has many of the cell-cell adhesion molecules of later species with more complex bodies. It has many of the transcription factor families of later animals with true tissues, but in the sponge these gene families have two to 34 times fewer members. Bona fide Hox genes are missing, although it does have more developmental transcription factors than the single-celled *Monosiga* described previously. The sponge also has many of the cell-signaling pathways of more complex species, but again, these gene families are impoverished in numbers by comparison to those of later animals.[20]

The sea anemone *Nematostella vectensis* is in the oldest animal phylum considered to have true tissues, Cnidaria, which also includes jellyfish and corals. These animals have among the simplest of body plans, with only two tissue layers and a saclike body that has a single digestive opening for both ingesting prey and excretion. A nerve net extends through the body, but there is no brain. The first draft of *Nematostella*'s genome predicts 18,000 protein coding genes, whereas another Cnidarian, the freshwater *Hydra*, is estimated to have about 20,000 genes.[21] Twenty-nine of 31 proteins used in cell junctions by later bilateral animals are already present in *Hydra*, showing the early potential for more complicated bodily architecture. Although there is no brain, there are neuromuscular junctions, but they are not as complex as in animals with more sophisticated nervous systems.[22]

The number of transcription factors in Cnidaria is greater than in the simpler sponges, but not as large as in more complex, later-evolving animals. A study of 10 major groups of transcription factors involved in bodily development (seven homeobox classes and three others) found that all the classes were already present in the last common ancestor of sponges and Cnidaria. But the number of genes increased in each class during the transition to the ancestor of Cnidarians and animals with bilateral body symmetry, and five of these ten classes increased further in the ancestor of insects and chordates (the protostome-deuterostome last common ancestor).[23]

An increase in the number of transcription factors also seems to have been important in the evolution of vertebrates. Amphioxus is a small fishlike filter feeder that lives most of its life buried in the sand on the sea floor. It is not a fish, however, or even a vertebrate, because it lacks true bones, although it has a stiff rod—known as a *notochord*—along its back for support. It shares many key features with vertebrates, however, including a hollow neural tube along its back (dorsal) side that serves as a central nervous system, opposing muscles groups that attach to its notochord to provide the undulating swimming motion characteristic of fish, and a postanal tail. These are characteristics of chordates, and vertebrates did not evolve until later when bones replaced the notochord for support. (In adult humans, all that remains of the notochord are disks between the bones of the spinal column.) Molecular studies now place amphioxus at the base of the chordate phylum, closest to the ancestor that would later lead to vertebrates, including ourselves.[24]

Amphioxus has been sequenced and is estimated to have about 21,900 protein-coding genes, a mere 2 percent less than the current number for humans.[25] Genetic analysis indicates there were two rounds of whole genome duplication of chromosomes between amphioxus and the origin of vertebrates, so that some genes on one chromosome in amphioxus, for example, are found on four chromosomes in humans. During the further course of vertebrate evolution, most of these quadrupled gene copies were lost, so the total gene count of humans is not very different from that of amphioxus. But particular families of gene duplicates were retained in the course of vertebrate history: those associated with cell signaling, transcription factors, neuronal activities, and developmental processes.[26] So the more complex bodies of vertebrates compared to amphioxus appear to be based not so much on the building blocks of structural proteins, but the architecture of how they are put together, as well as how cells communicate with each other. The instruments in the orchestra were established early, but the score of how they would play together became more complex.

The Cambrian Explosion

By the end of the early Cambrian, about 530 mya, a wide variety of animal body plans with bilateral symmetry is evident in the fossil record, and includes the forerunners of the insects and vertebrates.[27] It is generally agreed that single-celled choanoflagellates are the closest living relatives of multicellular animals. At some time between these single-celled ancestors and the appearance of the bilateral body plan, a great expansion of transcription factors occurred through large-scale gene duplications.[28] Several leading theorists in "evodevo," the study of evolution and development, suggest that new combinations of these transcription factors allowed structural genes to be put together in many new ways, facilitating the diverse new body plans that appear in the Cambrian explosion of new animal species.

Paleontologist Charles Marshall describes the period just before the Cambrian as a time when animals lacked a variety of sensory organs (e.g., compound eyes or antennae) and did not have appendages to interact with other animals or the environment, and there were no signs of predation except bore holes in the ground for possible escape. As these abilities began to appear in the Cambrian, they engendered arms races: advanced teeth, for example, appeared to promote stronger armor in prey species. Complexity begat complexity, and more peaks in the "fitness landscape" became possible. A developmental "tool kit" of genes that was already in place before the Cambrian appears to have facilitated this increase in complexity.[29]

Genes involved in early animal development have especially long upstream regulatory regions where many different transcription factors can bind. This allows them to integrate information from many neighboring cells. In contrast, ordinary "housekeeping genes" devoted to the basics of cell functioning have much shorter regulatory regions.[30] The genes for early development also produce messenger RNAs (mRNAs) with long trailer sequences where multiple microRNAs can bind, indicating an interaction between these two gene regulatory systems.

Evolutionary theorists Eric Davidson and Douglas Erwin propose that complex regulation of structural genes allowed more complex body plans to evolve from the Cambrian explosion onward. They describe regulatory networks that have a greater variety of subcircuits and more overall hierarchy in animals with a greater variety of cell types in their bodies.[31] We saw in figure 3.1 that the regulatory region of a gene is an information processor, combining a large variety of enhancing and silencing factors to determine how active a gene will be. The total structure, made up of many subunits, determines how active a gene will be in producing mRNA. The expansion of regulatory regions, and their interactions in many levels of hierarchical control, create a decision tree by which an animal's body plan is gradually worked out.

A particular transcription factor can be used over and over in different settings. Think, for example, how the number 7 can be used in a zip code (e.g., 78705) and also to specify an address within that zip code (e.g., 7214 Center Avenue). In an analogous way, a particular transcription factor can be part of the apparatus that turns on different genes at different stages of an animal's development. For example, the product of the *Ey* gene in the fruit fly *Drosophila* is used to regulate eye development, but is also required in the developing brain and central nervous system. A study of 67 transcription factors in *Drosophila* found that they helped control, on average, 124 other genes. The Twist transcription factor, which is involved in embryonic development, affects the expression of 500 target genes.[32]

The mixing and matching of transcription factors and the DNA sequences they bind allows for a great variety of control mechanisms for genes. The more sophisticated these control networks become, the more the potential for complex body plans. Only 3 to 5 percent of the total proteins an animal makes are transcription factors.[33] Thus a small percentage of genes could account for large changes in how other genes are expressed, and large changes in body plans could result from small changes in the total gene count. As Sean Carroll, a leading theorist in developmental biology writes, "Regulatory evolution creates new combinations of gene expression and therefore enables increases in the information content of genomes and the generative potential of development without expansion of gene number."[34]

MicroRNAs Provide Further Developmental Control

Not all RNA messages from the nucleus get translated into protein. In recent years we have found a special class of short RNAs, called *microRNAs*, that fine-tune gene expression. These have the ability to shut down protein production from the longer mRNAs. Every mRNA that exits the nucleus has both a leader and a trailer sequence around the actual message that codes for a protein. These untranslated regions have important signal sequences for how to use that mRNA, even though they do not code for parts of the actual protein. The trailer, for example, can contain a short sequence that is recognized and bound by a microRNA. If it is bound, that mRNA message is either degraded or inhibited from being translated into protein.[35] The cell generates such microRNAs when it wants to shut down production of particular proteins. Thus, while transcription factors can either enhance or silence output from a particular gene, microRNAs represent another level of control—and their effect is mainly inhibitory—when a cell is done utilizing a particular message.

The number of microRNAs a species produces roughly corresponds with body complexity. None have been found in yeast, the fruit fly has 171, the nematode worm *Caenorhabditis elegans* has 175, and 590 are found in mice and 940 in humans (see table 3.1).[36] Twenty-nine microRNA families were added at the base of animals with a bilateral body plan, another 41 were added at the origin of vertebrates, a further 63 with the evolution of mammals, and 84 novel families in primates. In the time that primates acquired 84 novel microRNAs, rodents acquired only 16 novel families.[37] Many of the new microRNA families in vertebrates are expressed in complex tissues such as brain and pharynx.[38] More than a third of human genes have target sequences for microRNAs, indicating the importance of this mechanism in human protein production.

MicroRNAs are highly conserved in evolution. Once they appear, very few changes are made in their sequences, and it is rare for descendent species to lose a microRNA family that appeared in an ancestor.[39] They are passed on like family heirlooms, and such evolutionary conservation is usually an indication of the importance of a molecule for survival.

Biologist Kevin Peterson and colleagues note that in three instances, rapid increases in the variety of bodily forms are associated with rapid increases in the number of microRNAs: the evolution of animals with three distinct tissue layers (triploblasts), vertebrates, and primates.[40] They suggest that the development of a species becomes more canalized and less "sloppy" over time as more of its protein-coding repertoire comes under the control of microRNAs. Increasing developmental precision may lead to more distinct types and more overall morphological complexity.

MicroRNAs can interact with the system of transcription factors already discussed. Genes concerned with embryonic development have especially long trailer sequences with many binding sites for microRNAs. This group of genes is among the most heavily targeted for microRNA control in terms of the total number of genes affected.[41] A large portion of early developmental genes are transcription factors and cell-signaling molecules. Early embryonic development often involves mapping out a building site before any structures are actually built on that site. That early mapping often involves accumulating different transcription factors at different sites. In the zebrafish embryo, for example, among 70 genes known to affect early body patterning, 34 percent encode transcription factors and 57 percent are signaling molecules.[42] Thus there is an intimate relationship between two systems that are important for early gene expression: transcription factors and microRNAs, and increases in both correlate with increases of morphological complexity.

The Brain as an Information System

Life gathers information in both genes and brains, and in both places microprocessors—mechanisms that allow the summation of multiple inputs to make a decision on the total—have evolved. Figure 3.1 shows how a variety of transcription factors can combine to affect the output from a gene. Figure 3.2 shows a neuron in the brain that is affected by the output of many neighboring neurons. To understand the similarity of the two systems, we should cover the basics of how a neuron fires.

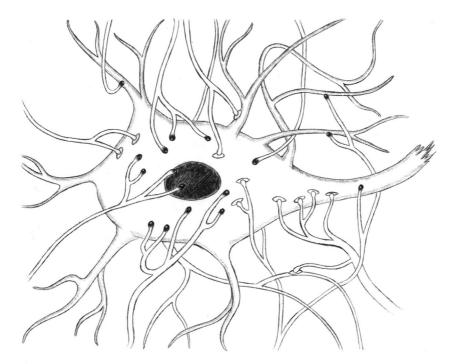

Figure 3.2 A neuron sums inputs from many neighboring neurons.
A brain neuron receives many inputs, some stimulatory (dark ends) and some inhibitory (light ends). It will sum all these influences and make a decision to fire or not fire. (From Neil A. Campbell and Jane B. Reece, *Biology*, 6th ed. [San Francisco: Benjamin Cummings, 2002], 1053. Adapted by permission of the publisher)

The end of a neuron is known as its *axon*. In figure 3.2, axons with dark tips represent stimulatory inputs and axons ending with light circles represent inhibitory inputs. All these inputs are impinging on a central brain cell, whose own axon is shown in part (cut off) going to the right.

A neuron at rest has a slight negative charge of about −70 millivolts (mV) in comparison to its environment. If it is raised to −55 mV, it fires, sending a spike of electricity down its axon. Figure 3.2 shows some of the many inputs a brain neuron receives, some inhibitory (making its interior more negative) and some stimulatory (making it more positive). What a neuron actually does is a sum total of all these inputs. If the total adds up to −55 mV or higher, it fires. If it is less than that magic number, it remains quiescent. Thus, like the regulatory region of a eukaryotic gene, neurons have become microprocessors, able to sum multiple influences and make a yes-no decision on the total. Genes, in fact, are a bit more subtle, because they can have different levels of activity, producing, for example, medium levels of mRNA versus high levels. Neurons only make a binary decision: to fire or not to fire.

We can compare the information content of these two systems to at least a first approximation. There are about 86 billion (86×10^9) neurons in the human brain,[43] and estimates of the average number of connections (synapses) between them vary widely, from about 500 up to 10,000.[44] I will use a conservative number of 1,000 (10^3). Each input will fire or not fire, making a binary decision, so we can estimate the total information capacity of the brain as

$$86 \times 10^9 \times 10^3 = 10^{12} \text{ or 86 trillion bits.}$$

There are about 22,300 genes in the human genome, covering about 3.2 billion (3.2×10^9) bases of DNA. Structural genes cover only about 1.5 percent of this expanse, but another 3.5 percent is thought to regulate gene expression.[45] (This regulatory distance is large enough to encompass transcription factors as well as other ways in which DNA is modified, such as epigenetic factors.) Genes average about 2.6 alternatively spliced forms. There are four bases of DNA (A, T, G, C), so each could be represented by two bits of information (00, 01, 11, 10). Putting all these data together, we can calculate the information content of the human genome as

$$3.2 \times 10^9 \times .05 \times 2.6 \times 2 = 83.2 \times 10^7 \text{ or 832 million bits.}$$

These numbers are an approximation, but they indicate that the information content of the human brain is many orders of magnitude greater than that of the genes in the human genome. In both systems, the combinatorial possibilities are nearly infinite. That is, the number of ways 22,300 genes might interact, or the circuits that could be formed by 86 billion neurons, are nearly endless in their possibilities. But these numbers suggest that the nervous system may have gone beyond the genes in its information capacity, and there is physical evidence for this in the way the brain develops.

Carl Sagan did a similar analysis and pointed out that during animal evolution, there came a point when the information capacity of brains exceeded that of the genes alone.[46] He also noted that humans have extended this long-term trend with the storage of cultural information in books, computers, and so forth. Human culture is not just a peculiar adaptation of a language-specialized primate. We are extending one of the basic trends of evolution, which is to accumulate information as a way to adjust to fluctuations in the environment. Nervous systems can do this faster than the genes by themselves and have extended the range and speed by which an organism can respond to change.

With denotative language, humans further increased the pool of information that could be shared between individuals. The invention of writing enabled this knowledge to be transferred across generations without relying on just the memories of living people who could speak to each other. Computers and the Internet have extended this trend further, approaching the point where all knowledge could be available to everyone at all times, forming a kind of global brain.[47] In 2009, the indexable Web was estimated to contain at least 25.2 billion pages, far exceeding that of the books in the Library of Congress.[48]

The accumulation of knowledge has allowed a great mastery of nature, new ways to use materials, gain energy, secure food, and fight disease. It can be seen as a continuation of the information gathering that began in genes and advanced in brains, as a way to store data about the environment and find ways to respond to change. The large-brained mammals and birds are more homeostatic than groups that came before them, able to be active in a wider range of conditions. We humans, who can be considered the dominant animal on Earth today, have capitalized on this trend. We seek to secure and extend the human enterprise, as each species seeks for itself (although this is a result of the self-interest of each individual working for himself or herself).

Plants, which do not have nervous systems, must encode almost all their behavior in genes, and this may explain why they have higher gene counts than animals. An animal can reach for nourishment; a plant has to grow toward it, and to do that it must marshal all the genes involved in cell proliferation. An animal can run away from a predator; a plant must activate predator defenses, making toxins to dissuade herbivory or walling off an area damaged by predation. Animals have courtship rituals; plants draw attention by producing showy flowers. At almost every stage of the life cycle, actions accomplished via nerves and muscles by animals must be achieved with gene expression by plants. Brains are not present to supplement gene activity, and therefore it may not be surprising that some plants have higher total gene counts than animals.

This may also explain the unusually high gene counts in some invertebrate animals. Species with high levels of instinctive, fixed-action behavior may have more activities encoded in their genes than animals with higher levels of flexibility and learning. The way in which the brain can go beyond simple genetic determinism can be seen in how synapses are formed in some large-brained species.

Wiring by Firing

Genes alone do not determine the circuits of the human brain; experience has a lot to do with how neurons are finally connected to each other.[49] Humans have the greatest number of synapses at about two years of age, and a pruning process goes on until sexual maturity. Some areas will lose as much as 50 percent of their connections, and different brain areas mature at different times.[50]

The process has been well studied in the visual cortex, an area at the back of the brain that maps input from the eyes. Early in development, each brain neuron may receive input from both the left and right eyes. But because the two eyes on either side of the nose see objects from slightly different angles, their firing pattern will be somewhat different. A brain neuron dedicated to interpreting vision will fire in synchrony with one or the other input from the eyes. It will also release growth factors in limited amounts to the neurons that stimulated it. The result is that connections that fire in synchrony will be strengthened, while those that are not correlated will be pruned back (figure 3.3). Neurons that fire together, wire together.

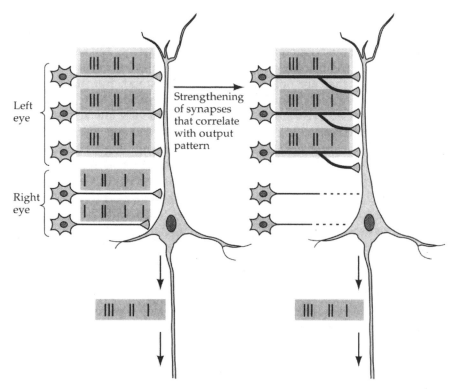

Figure 3.3 Wiring by firing.
A brain neuron receives input from both the left and right eyes early in development. It begins to fire in response to the pattern of stimulation from the left eye (patterns of firing are represented by short vertical lines). Connections between synchronized neurons are strengthened, whereas connections to the uncorrelated right eye wither away (*right*). (From Dale Purves et al., eds., *Neuroscience*, 3rd ed. [Sunderland, Mass.: Sinauer, 2004], 570. Reproduced by permission of the publisher)

Normally the mature visual cortex has alternating columns of cells, some possessed by input from the left eye and others dominated by the right eye. The role of experience in this process is illustrated by experiments with cats. If a patch is put over one eye of a young kitten and left there until adulthood, it will become blind in that eye, for there will be hardly any brain neurons that respond to it anymore. The cells in the visual cortex will be entirely taken over by input from the other eye.

No such effect happens if pirate patches are put on mature cats, illustrating an early "critical period" when brain connections are made.[51]

Puberty appears to conclude a major period of wiring in the human brain. A number of human characteristics are set at this point, such as language ability.[52] If you learn a language as a child, for example, you can speak it like a native speaker. If you learn it after puberty, it is likely you'll have an accent that forever brands you as a foreigner. People can go on learning their whole lives, of course, but this appears to be more a matter of "soft-wiring," of modifying existing connections, rather than the "hard-wiring" of establishing those connections in the first place. For similar reasons, the top musicians, as well as champion athletes in sports like golf and tennis, all began their game as young children.

In the brain, information gathering has gone beyond the activity of the genes alone. DNA builds an excess of structures that are relatively unspecified in their connections, and experience shapes and prunes the final patterns. The genes build a kind of photographic plate that slowly develops as it is exposed to the light of experience. Both nature and nurture, therefore, influence how our brains develop. This belies ideas of simple genetic determinism in behavior. Experience has at least as much influence on the connections between neurons as the pathways set up by genes do. Thus the information content of the final architecture of brain wiring can be greater than the information stored in the genes alone. The calculation of the greater bit content of the brain than of the genes really shows something about how these systems are constructed.

The process of wiring the brain has sometimes been called *neuronal selection*, and it resembles clonal selection in the immune system, as discussed earlier, and natural selection in populations of organisms. In all three cases, excess is created, often by a random process that produces great variation. The brain begins with more synapses than it will ever use, the immune system produces more B cells than needed to fight diseases, and in natural populations sexual recombination produces a greater variety of offspring than resources can usually support. A competition ensues for those variants best suited for current conditions, and those individuals are strengthened or multiplied. In all three cases, nature cannot predict the needs of the future, so it deals a lot of cards from the deck in order to find a winning hand. Randomness has a part in each process, but it is randomness used for a purpose. It is a way of probing the unknown space ahead, of seeking the best path in a dark wood, and then reinforcing the variants that are most successful.

Encephalization Quotient

A process of "encephalization" has taken place over evolutionary time, from the diffuse nerve net of a jellyfish to the concentration of neurons in a central nervous system, as found in vertebrates. In fish, the brain is a fairly linear structure, with nearly equal portions of forebrain, mid-brain, and hindbrain, whereas in later evolution, increasing executive control is localized in the forebrain.

Biologist Harry Jerison analyzed brain and body data for 198 verte-brate species and found, on average, a 10-fold increase in brain:body ratio for mammals and birds over that of fish and reptiles.[53] A later study ex-panded this data set to nearly 2,000 species and found similar, although not quite as great, differences in relative brain size between vertebrate classes.[54] Figure 3.4 shows minimum polygons that can be drawn around the data points for 198 vertebrate species and indicates the magnitude of the brain-body size differences between them.

It is generally agreed that brain weight should be looked at in reference to body weight. If the brain is a kind of control center for the body, it makes sense that it should get bigger when it has more tissue to control. Humans do not have the largest brains: elephants and whales are bigger in this regard, but so are their bodies. Comparing brain weight to body weight in a ratio gives a more realistic sense of relative brain size for a species.

But there are problems in too simple a comparison. Brain weight in-creases more slowly than body weight as species become larger, so that a mouse has a larger proportion of brain than an elephant. In a pocket mouse, brain is 10 percent of body weight, in humans it is 2 percent, and in a blue whale it is 0.01 percent.[55]

A better comparison looks at an average-size brain for a particular body weight and asks whether a species has a larger or smaller brain for that body size. But here too there are problems. Depending on the data set, the slope of the average regression line through many species will vary. Jerison found a slope of ⅔, whereas John Eisenberg, with a larger data set, found ¾.[56] Different taxa have different average lines: the slope for mammals is steeper than the one for birds or reptiles. Some research-ers have suggested that the line should curve downward as a quadratic function.[57]

Whatever the regression line, we can still compare species at a given body weight in terms of their relative position. Encephalization quotient

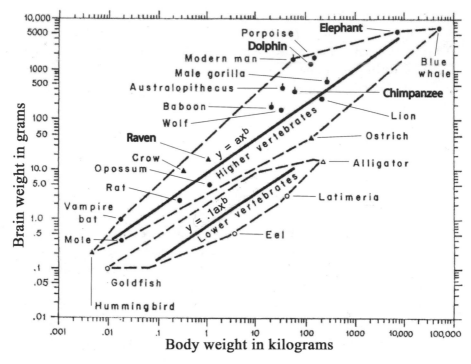

Figure 3.4 Minimum polygons drawn around brain and body data for 198 vertebrate species.
Lines through the center of each polygon represent average brain size for animals at a particular body size. Animals above this line have an EQ greater than 1, and animals below it have an EQ less than 1. Data were transformed to logarithms to accommodate a wide range of values. (From H. J. Jerison, "Brain Evolution and Dinosaur Brains," *American Naturalist* 103 [1969]: 575–588. Adapted by permission of the University of Chicago Press)

(EQ) is a measure of the size of an animal's brain for a given body size. An animal with an EQ of 1 has an average brain weight for animals of its body weight. Numbers above 1 indicate a brain larger than average, and numbers below 1 indicate a smaller brain. EQ values for a variety of species follow this paragraph, showing, where possible, the range of sizes found within a species. The numbers are taken from zoologist Eisenberg's large data set.[58] Humans have by far the largest EQ of any species, although primates in general have higher EQ values than most other taxa. Also, the animals nearest to us just in terms of relative brain size are actually porpoises and dolphins:

Human: 7.69–7.33

Chimpanzee: 2.55–2.18

Gorilla: 1.68–1.4

Rhesus macaque: 2.34

Ring tail lemur: 1.45

Common porpoise: 4.9

Dolphin: 3.23

Pilot whale: 1.7

African elephant: 2.05 (female) / 0.97 (male)

Asian elephant 1.72–1.62

Horse: 1.07

Domestic cattle: 1.09

Wild boar: 0.27

Fox: 1.89

Gray wolf: 1.33 (male) / 0.92 (female)

North American raccoon: 1.5–1.37

Tiger: 1.35

Lion: 0.83–0.57

Eastern chipmunk: 1.74

House mouse: 0.81

Norway rat: 0.79

Marsupials:

Common opossum: 0.573–0.353

Australian native cat: 1.05

Tasmanian devil: 0.42–0.36

In chapters 5 and 6, I will examine some of the species with the highest EQs and show that they have a surprisingly human constellation of qualities. These species are listed in boldface in figure 3.4. They all have complex societies and communication and manipulate their environments in intricate ways. They are all K-selected, equilibrium species that have slow development and long lives, with an emphasis on learning and behavioral flexibility. There may be common qualities associated with large relative brain size.

High EQ is also a measure of the "extra neurons" in an animal's brain.[59] If an average animal can survive with an EQ of 1, an animal with an EQ of 2 or 3 may have many extra neurons that can be devoted to functions beyond just regulating the body and surviving in its niche. We will see, for example, that higher-level abstractions in social life and tool making are associated with high-EQ species. Humans, with nearly twice the EQ of the next nearest species, create abstractions and symbols far beyond that of any other animal.

Some researchers have suggested that absolute brain size rather than relative measures like EQ are a better index of cognitive abilities in a species.[60] A chimpanzee has a total brain mass in grams about 28 times that of a raven, yet in spite of this difference in absolute brain size, the two share many of the same mental abilities, as will be discussed in chapter 5. Any index that ignores the fact that in general organs scale up with increasing body size will miss important convergences of behavior that take place between species of very different brain and body sizes.

The polygons in figure 3.4 show that relative brain size has increased over evolutionary time. Dinosaurs generally would fall within the polygon of living reptiles in this figure, although a few had larger brain:body ratios.[61] The earliest mammals, from about 225 mya, had brains about four times larger than their reptilian contemporaries. Jerison speculates that they occupied mainly nocturnal niches and their increased brain matter was necessary for processing input from heightened senses of smell and hearing that life at night required.[62] This was a period of prolonged climate warmth, when mammalian adaptations such as homeothermy (being warm-blooded) would not have been an advantage, and when large reptilian carnivores already roamed the planet.[63] These early nocturnal mammals were no larger than a house cat and were probably insectivores. Mammals did not increase further in relative brain size until after the demise of the dinosaurs, and only in the last 55 million years have they reached their nearly 10-fold increase in brain size.

By 20 mya, birds had achieved modern brain sizes.[64] Zoologist A. Portmann did an extensive anatomical study of birds and also found that they have increased an average of 10-fold in brain:body ratios compared to reptiles. In birds as well as mammals, many learning skills are encoded in the cerebral hemispheres of the forebrain. Portmann found some of the largest hemisphere sizes in songbirds, with their highest values among the corvids: the ravens, crows, and jays.[65] We will examine some of their remarkable behavior in chapter 5.

Brain Development

How can some 22,300 genes code for the connections between 86 billion neurons in the human brain? And why does a human, with about nine times the brain:body ratio of a mouse, have about the same number of genes?

The answer appears to be that the genes are building an open-ended mechanism. They lay the groundwork but do not specify the final connections. Experience will also be important for determining the final circuits, which is one reason we should look beyond mere genetic explanations for behavior.

The nervous system of all vertebrates begins as a hollow tube that runs from head to posterior. In humans, at 25 days after conception, there is a swelling at the front end that is the precursor of the brain. By

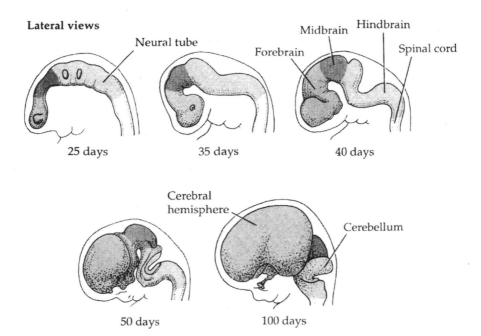

Figure 3.5 Development of the human brain.
The brain begins as a swelling at the front end of the neural tube that runs along the back of the embryo. The foremost part of this swelling becomes the forebrain, which by 100 days mainly consists of two cerebral hemispheres. The cerebellum grows from the hindbrain. (From Dale Purves et al., *Life*, 4th ed. [Sunderland, Mass.: Sinauer, 1995], 854. Reproduced by permission of the publisher)

40 days, the front part has developed two lobes that are the beginnings of the cerebral hemispheres.

The hallmark of mammalian brain development, and especially human development, is the enormous expansion of this area of the brain (figure 3.5). In humans, the cerebral hemispheres continue to grow until like a helmet, they cover most of the rest of the brain.

Below the cerebral hemispheres, growing out from the hindbrain, is the cerebellum. It too will develop twin lobes and sit as a "little brain" below the larger cerebral hemispheres. It maintains reciprocal connections with frontal areas of the cerebral hemispheres and is involved in the integration of sensory and motor functions, visual problem solving, and the learning of procedures. It has evolved in concert with the upper cerebral cortex; the two areas vary together in numbers of neurons across four orders of mammals, comprising 19 species, including humans.[66]

The uppermost area, or cortex, of the cerebral hemispheres changes the most in volume when you compare small-brained animals to large-brained species. In humans, as well as in dolphins and elephants, this area has many folds and fissures to accommodate its large surface. It has been estimated that if our cortex were unfolded and fully expanded, our skulls would have to be the size of beach balls to contain it. Most of this surface area has six layers of neurons and is sometimes called the *neocortex* to distinguish it from areas with fewer layers, as are found mainly in earlier evolving groups like reptiles and amphibians. The neocortex also changes the most between mammalian species. In humans it makes up 96 percent of the brain's cortical surface; in hedgehogs, it is only 32 percent.[67]

All the cells that make up the six layers of the cortex have their origin in the hollow nerve tube from which the brain grows. A set of precursor cells near the edge of the hollow interior divides repeatedly and gives rise to neurons that migrate upward and form the swellings that become the brain. It does not take many extra cell divisions to build a large brain instead of a small one. Cells always divide in two: therefore two cells make four, four make eight, eight make 16, and so on. Only 17 doublings explain the brain size difference between a shrew and a human being.[68] This is probably why we do not find much difference in the number of genes it takes to build a large-brained animal or a small-brained one. All that is needed is a few changes in regulatory genes that tell precursor cells to divide more often. This is the most likely reason that it does not take more genes to build the brain of a human than it does that of a mouse.

Early brain growth in humans produces about two times more neurons than will be present in the adult brain. Brain cells that do not receive the proper stimulation die away in a programmed manner known as *apoptosis*. Again, changes in a few regulatory genes, either increasing cell divisions or decreasing the amount of neuronal death, could account for large changes in brain volume.[69] Neurobiologist Pasko Rakic calculates that only four extra rounds of cell division near the cortex of the human brain would result in a tenfold increase in surface area.[70] Humans achieve their large brain size by an extension of fetal growth patterns for the brain. In most primates, the rate of adding new neurons levels off at birth. As figure 3.6*a* shows, the fetal rate of brain growth in humans continues well after birth and will not level off until about 21

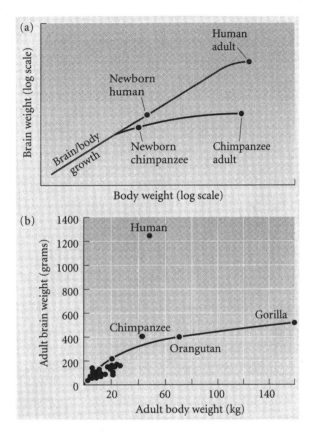

Figure 3.6 Brain growth in humans compared to other primates. (*a*) Humans continue fetal growth rates of brain neurons until 21 months, leading to (*b*) brain:body weight ratios about 3.5 times greater than those of apes. (From S. F. Gilbert, *Developmental Biology*, 8th ed. [Sunderland, Mass.: Sinauer, 2006], 391. Reproduced by permission of the publisher)

months. We emerge untimely from the womb and are essentially extra-uterine fetuses for about a year. This is the result of our unusually large skulls at birth, which can barely pass through a woman's pelvic bones, making human delivery unusually painful, prolonged, and dangerous compared to that of other mammals.

After birth, we are still adding about 250,000 neurons per minute, and rapid brain growth will continue for about two more years.[71] In terms of mass, this leads to a very high brain:body ratio compared to other primates, as figure 3.6*b* shows. Recent research, however, indicates that in terms of total cell numbers, humans have about as many brain neurons as expected for a primate with our brain size.[72]

The chimpanzee genome has been sequenced, and only 1.23 percent of our DNA differs from that of our near living cousin.[73] Yet we have brains roughly three times the size of a chimp's in brain:body ratio. Results show large changes in small sets of genes, especially those that regulate the activities of other groups of genes. In humans there has also been rapid evolution of a group of genes related to brain size and behavior.[74] Six of these genes, for example, cause severe reductions in human brain size (microcephaly) if they become defective. There are also patterns of gene expression that are unique to the human brain, indicating that we may also have some unique circuits there.[75]

We can think of humans as primates that have extended several youthful characteristics during development. We remain hairless, like most newborn primates. We retain high brain:body size, and are innovative and playful—all characteristics of young animals. This prolongation of juvenile characteristics begins with the extension of fetal rates of brain growth and may be due mainly to changes in small sets of genes that regulate the rate of growth. As we saw with changes in body complexity, alterations in small numbers of regulatory genes rather than large-scale alterations in gene number are the most likely explanation for increases in overall complexity. Brain wiring is also open to the input of experience, adding to the intricacy of its final architecture.

The Conservation of Information

Eukaryotes are more conserving of information than prokaryotes. Eukaryotes build their proteins in sections known as *exons*. Often an exon is a functional unit of a protein, much as the handle and head of a hammer can be thought of as different functional units. A handle might be combined with different kinds of head, such as that of an axe or a gauge. In a similar way, eukaryotes have evolved ways of mixing and matching functional domains in order to create new proteins. Thousands of proteins can be explained by hundreds of exons combined in new ways. Sections of DNA known as *transposons* can jump from one area of a chromosome to another and carry exons, or other useful parts of a gene, with them. Binding sites for transcription factors can be mixed and matched by similar mechanisms to create great variety in how genes are controlled.

Eukaryotes therefore conserve the information in useful domains and tinker with new ways to combine them, much as the parts of a Tinkertoy set can be put together to produce a wide variety of structures.

Prokaryotes rely much more on raw mutation to evolve new genes and new control areas of genes. They do not build their proteins out of exon subunits. Ninety-two protein families in eukaryotes are made by combinations of domains that appear separately in different proteins made by prokaryotes.[76] Different forms of a protein can also be expressed in a eukaryote by splicing together different sets of its exons. Ninety-five to 100 percent of human genes are believed to have alternatively spliced proteins of this nature. For some genes, thousands of alternative forms are possible, using different splice sites. The number of alternative splices appears to be much higher in mammals than in nematodes or flies, although this subject has not yet been thoroughly studied.[77] Neurons in mammals appear to be especially rich in the number of alternative protein forms they can produce. Domain shuffling is more frequent in multicellular animals than in single-celled eukaryotes, and unique domain combinations were important in the evolution of vertebrates. For example, cartilage, a unique vertebrate protein involved in skeletal support, was created by combining domains from proteins that did not originally have a skeletal function in nonvertebrate ancestors.[78]

The Value of the Individual

The conservation of information also applies to sex. Eukaryotes in general are much more selective than prokaryotes about the genetic information they will incorporate into themselves. Bacteria have mechanisms to exchange segments of DNA, but they also regularly accept sequences brought in through viral infection or that they take up from their surroundings (from other bacteria that have died and broken apart). If one of these sequences is similar to one of the bacterium's own genes, it can cross over and replace the original gene on the bacterial chromosome.

Eukaryotes rely more on sex for new combinations of genes than prokaryotes do, and this imposes a screen on possible partners. The ultimate source of all variation in genes for both eukaryotes and prokaryotes is mutation. But sexuality imposes a screen on this variation. Just making it to reproductive age is, after all, a severe test of the hardiness of an individual's genes, and therefore accepting genes from a mature partner means they have already gone through many tests of fitness. Moreover, sexual attractiveness has been shown to have much to do with health and vigor.[79]

There has been something like an evolutionary increase in the selectivity of mating choices in eukaryotes. Among marine invertebrates that evolved early, sex is relatively anonymous. Corals and many mollusks shed large numbers of sperm into the sea, and these can fertilize any female of the species nearby. Among birds and mammals, there are often elaborate courtship rituals in which one individual makes complex choices about the suitability of a mate. Studies have shown that these are often choices for "good genes." Bright colors that attract female birds, for example, are an indication of the male's health and freedom from parasites. Female mice seek greater diversity for their immune systems, something they can actually tell from the odor of the markings a male leaves behind.

The biological process of forming eggs and sperm, known as *meiosis*, has a random component. We do not get to choose which of our genes gets put into one of our reproductive cells. I would love to be able to put only the best of myself (whatever that might be) into my children. But we do not get that option. However, we do have some choice over our partners, and because reproduction usually comes at maturity, those genes have already stood the test of time. Meiosis, therefore, imposes randomness on tested genes and is a step removed from the randomness of raw mutation. It creates variations on a proven theme.

The increasing selectivity of courtship in large-brained animals is part of the increasing conservation of information in evolution. Over evolutionary time, the information content of genes and brains, and the importance of the individual as a repository of useful information, have both grown. The complexity of genomes and relative brain size have also increased, with the genes eventually laying the groundwork for relatively open-ended nervous systems that appear to have more information content than the genes themselves do. Microprocessors have evolved in both genes and neurons that can perform summations of multiple influences. New proteins evolve not just by simple mutation, but through combinations of domains (exons) that have proved to be useful. And information is conserved through the elaboration of courtship that allows the individual some choice over the genes incorporated into the next generation.

As the information content of an organism increases, it has more possible combinations of parts and becomes more distinctly individual. We will see that species with large relative brain size have great variability of behavior and more complex communication and social organization, and individuals appear to value each other highly in terms of long-lasting relationships and helping behavior. These qualities of large-brained animals will be explored in detail in the next few chapters.

4

What Is a Big Brain Good For?

The brain:body ratios of mammals and birds is about 10 times greater than that of reptiles, amphibians, and fish. In any other area of biology, such a difference would be considered significant: a digestive tract that was several times longer or a comparably larger wing or muscle would indicate a change in abilities and prompt a search for the evolutionary pressures that made it possible. Yet in some areas of cognitive biology, it is claimed that large brains make no difference at all in the learning abilities of different species.[1] In part this is due to the difficulties of designing tests of "intelligence." We are aware of cultural biases in IQ tests for humans; how much greater must this problem be if we are trying to compare different species, such as pigeons, rats, and monkeys? Each of them may rely on different sensory modalities so that, for example, rats do poorly on visual tests but excel where odor is an important cue. It is probably rightly claimed that there can be no unitary definition of intelligence.

But the brain is an expensive organ.[2] In humans, it constitutes about 2 percent of body weight but consumes 20 percent of total energy intake. It flies in the face of basic evolutionary theory that any animal would incur such high costs without accompanying benefits. Although lab tests of learning generally have been frustratingly inconclusive, field studies in recent years have shown significant differences in social complexity, innovation, and behavioral versatility for species with high relative brain size. The rate of evolution itself may be affected by large brain size and the behavioral flexibility it affords.

The Social Brain

Anyone who has survived the rigors of high school has a sense of how complicated social life can be. There are cliques, enemies, and alliances, and considerable energy goes into dealing with relationships. Females are often better at these things, and everyone is familiar with the "nerd" who has some great technical skill but is clueless about what makes relationships work. The social world of some large-brained animals apparently is not too different. Primatologist Frans De Waal has written extensively about "chimpanzee politics."[3] Although many nonprimate species live in simple dominance hierarchies, chimpanzees have complex relationships in which coalitions, often with nonrelatives, decide rank. Alliances can shift over time, and two lower-ranked males, for example, may combine to overcome an alpha male. Female support may also be essential for a male to retain his leadership position.

Mediated reconciliation, often performed by females, has also been observed in chimp colonies. After a fight, 20 instances have been recorded where a female groomed one male, led him to the second male, groomed his opponent, and then left them to groom each other.[4] An understanding of tertiary relationships, more complex than just self and one other, is evident in all the large-brained species we will look at in detail in the next two chapters.

The size of grooming cliques has been positively correlated with forebrain size in 30 species of primates, including humans, where barbershop gossip still gives a sense of what goes on during this primal activity. Relative brain size is a predictor of a variety of social activities in primates, including social group size and complexity, frequency of coalitions, social play, tactical deception, and social learning (the cultural transmission of skills).[5] A comprehensive study of 206 species of mammals in three orders found a significant relationship between relative brain size and sociality.[6] (Throughout these chapters, *relative brain size* designates a measure similar to encephalization quotient [EQ]. It measures how much a species' brain differs from the brain size of an average animal of its body size.)

It does not take much brain power to follow the herd, but a long-term relationship with single individuals can be very demanding, as many a married couple will attest. In carnivores, ungulates (hoofed animals like deer and sheep), birds, and bats—species that are pair-bonded—have relatively larger brains than species with other mating systems.[7] Many birds are monogamous, but there can be a difference in the quality of relationships here also. Geese mate for life but have prolonged physical contact only during the mating season. Rooks also mate for life, but year-round they continue food sharing, bill twining, mutual grooming, and aiding each other in fights with neighbors. They are part of a family of birds known as *corvids*, which have unusually high brain:body ratios. Geese, with smaller relative brain size, build simple nests on the ground, and their young can feed themselves after hatching. Rooks build more complex nests together and mutually feed their young for a long time after they hatch.[8]

The politics of group life may also require deception. Primatologist Robin Dunbar notes "grade shifts" in forebrain size in going from prosimians to monkeys to apes. The shifts in size correlate with the amount of tactical deception in social relationships. This activity, in which one individual seems to deliberately mislead another to the actor's advantage, has been described as "Machiavellian intelligence" and is the subject of many animal studies.[9] A survey of 18 species of primates living in the wild found that forebrain size predicted how much individuals used deceptive tactics for social manipulation.[10]

Much social communication takes place in primates via facial expression. Monkeys and apes are unusual in both the number of color patterns on the face and the number of muscles that control facial expression.[11] In general, great apes and humans have more complex sets of facial muscles than monkeys and prosimians. The leaf-eating gorilla has about as many facial muscles as the omnivorous chimpanzee but displays a relatively impoverished range of facial expressions.[12]

To see how much humans rely on just facial expression for communication, try turning down the sound on a television soap opera some time and just watch peoples' faces. In most cases you can guess their moods and how they feel about each other without hearing any words. A large part of the motor area of the human cortex is devoted to controlling the unusually high number of muscle pairs we have for facial expression, and distinct areas of our brains interpret the facial expressions of others. This mode of communication may be important to primates in general. An experiment with rhesus monkeys in which the nerve that controls facial expression was cut found that when these individuals were returned to their troop, they soon dropped in dominance rank and were treated more aggressively.[13]

Diet

In some families of mammals, brain size is related to diet, with food sources that are unpredictable in terms of space and time apparently making greater demands on the brain. A study of seven families of primates found that groups living on fruit had larger relative brain sizes than those living on leaves.[14] Brain size was also related to home range size because leaf eaters (folivores) can live on local supplies whereas fruit eaters (frugivores) have to remember a larger territory where different plants come briefly into fruition in widely separate places. Because it generally takes a longer gut to digest leaves than it does to digest fruit, it might also be that leaf eaters simply have bigger bodies, giving them a small brain:body ratio. But when researchers compared body length instead of body weight to brain weight, the distinction remained, indicating that a diet of leaves really does correlate with a smaller brain. Great apes showed the highest relative brain size in this study, while lemurs were lowest.

A study of small mammals (rodents, insectivores, and members of the rabbit family) found similar results using the same statistical methods.[15] Leaf eaters had smaller brain:body ratios, and this distinction remained when body length rather than body weight was used in an effort to compensate for possible differences in gut size. A study of bats found that species that lived on energy-rich food that was distributed unpredictably in space and time, such as frugivores and nectarivores, had larger brain:body ratios than aerial insectivores that fed on local supplies using echolocation.[16]

How much harder can it be to eat fruit than leaves? Tropical forests are much more diverse and patchy than northern forests, and fruit on trees generally is available for shorter periods than young leaves. Primatologist Katharine Milton compared the lifestyles of two species of neotropical monkeys that have about the same adult size and occupy about the same geography to illustrate the differences between a frugivore and a folivore.[17] Howler monkeys (*Alouatta palliata*) live mainly on leaves and have an average home range of 31 hectares (77 acres), whereas spider monkeys (*Ateles geoffroyi*) live mainly on fruit and have a home range of about 800 hectares (about 1,980 acres). On an average day, spider monkeys have to travel three times farther for a meal. At six months old, a leaf-eating howler is relatively independent of its mother, whereas at 24 months a young spider monkey is still being carried by its mother and nursing. Spider monkeys have about twice the brain:body ratio of howlers, as well as a more complex repertoire of facial expressions, vocalizations, and engage in elaborate bouts of grooming each other. Both species are very social, but howlers do almost everything together whereas spider monkeys have what is called a *fission-fusion* society: the troop splits up into smaller groups to find food in different locations and then rejoin later in the day. The patchy distribution of fruiting trees makes this strategy necessary.

Milton notes that fission-fusion is the social organization of several species that are at peaks of brain development: chimpanzees, dolphins, and parrots. All have to learn the location of widely disparate food sources and all have long dependence on parental care, complex societies and communication, and high brain:body ratios. All these life history characteristics may be involved with each other, and also in the evolution of human beings, who spent millennia as hunter-gatherers searching for sparse resources before settling down to agricultural life.[18]

Innovation and Tool Use

Biologist Louis Lefebvre and colleagues culled from journals thousands of reports on innovative behaviors of birds and primates in the wild.[19] They controlled statistically for factors that are reported more often in wildlife journals, such as colorful species. They found a positive correlation between innovation rate and forebrain size in both groups. In both primates and birds, relative brain size correlates with technical innovation and the acquisition of new food sources.[20] The avian brain has a very different architecture from the primate brain, although both have expansions of the forebrain that are implicated in higher associative functions. Large forebrain size, rather than any common ancestry or ecology, is the best explanation for this convergence of abilities.

In both primates and birds there is a statistically significant correlation between innovation rates, social learning, and tool use. Researchers distinguished between "true tools," such as hammers that are detached from their source and held by the animal, and "borderline tools," such as stones or pavement against which things can be thrown—that is, implements that are not directly manipulated. They found that true tool users have a larger overall brain size than prototool users. Prototool use also involves more stereotyped and repetitive actions than do the complex manipulations of true tools.[21] In a review of 125 cases of tool use among 104 species of birds, the most cases were found among the large-brained corvids, with the common crow having the largest repertoire of tools and techniques.[22] All four species studied in the next two chapters are tool users.

Invasion Success and Evolution

Innovation is apparently related to success in new environments. Biologist Daniel Sol and colleagues studied the 100-plus bird species that human immigrants have imported to New Zealand.[23] They found a link between forebrain size, feeding innovations, and invasion success. Successful invaders also had a history of more feeding innovations in the country they originally came from. This study was widened to include 645 introductions of 195 bird species around the world. Again, there was a positive correlation between relative brain size and the ability to become established in a novel environment.[24]

Resident species of birds have larger relative brain size than migratory species that leave during the winter.[25] It has been suggested that species that stay on through the cold have more flexible behavior and can adjust to a wider range of conditions. Resident species increase their feeding innovations in winter, whereas migratory species do not have to change their diet as much.[26]

The flexibility to deal with seasonal change may also affect the ability to survive overall habitat change. British birds are among the most thoroughly studied in the world, and the English landscape has been extensively modified by farming and urban growth over the years. A study found that large-brained avian species have declined less than those with smaller relative brain size.[27] Similar results come from a study of 99 species of neotropical parrots distributed across Central and South America. Researchers rated variability of the environment in terms of temperature and rainfall, spatial heterogeneity, and habitat type. They found that large-brained species could tolerate a wider range of conditions than those with smaller relative brain size.[28]

Similar results were found in a study of birds that had the "street-smarts" to flourish in urban environments.[29] Researchers compared 22 avian families in their ability to colonize and breed within 12 European cities. They found that successful species had higher brain:body ratios than those that avoided urban life. Researchers attributed the success of these species to their ability to innovate and adapt to changing environmental conditions.

Similar patterns can be found over a longer time frame. Fossil evidence indicates that parrots had already achieved large brain size before the great extinction event that wiped out the dinosaurs. Their survival in niches after the extinction has been attributed by some researchers to their behavioral flexibility and tolerance of a wide range of ecological conditions. The ability to adopt innovative ways of feeding may have been key to survival in a radically changed world.[30]

Flexible behavior may thus have evolutionary consequences. J. S. Wyles and colleagues measured eight anatomical traits representing all major body parts on the skeletons of 239 bird species and subspecies.[31] By comparing the amount of anatomical diversity in each group with the group's time of emergence in the fossil record, they obtained an estimate of the rate of anatomical change. Birds in general have a high rate of change, and songbirds have a rate more than twice as fast as that of other birds.

Combining these data with 20,000 measurements from other animal taxa, they derived an estimate of the rate of anatomical change in relation to brain size.[32] Birds and mammals have a rate nearly three times that of lizards and frogs, and the lineage leading to humans has the highest rate of all. The authors relate these changes to behavioral flexibility. They suggest that large brains allow high rates of innovation in feeding strategies and the exploration of new niches. New adaptive strategies could then be subject to natural selection for physical traits that enhance those strategies. In this way, behavior might sometimes be the leading edge of evolutionary change, as some theorists have suggested.[33]

After the demise of the dinosaurs, mammals rapidly diversified and increased in brain size two to four times (as measured in EQ values). Orders of mammals that went extinct had smaller EQ values than those that survive today, indicating that there was positive selection for greater brain size.[34] This was a period when placental mammals diversified into many new niches, and it is perhaps their behavioral flexibility that allowed them to explore new opportunities. This period of the early Eocene also saw a rapid rise in atmospheric oxygen that facilitated the high metabolic needs of placental mammals. A fetus sequestered inside the mother must exist at relatively low oxygen (O_2) levels, and this is possible only when the mother can breathe high levels of atmospheric oxygen.[35]

Lab Studies

Evaluating a large number of reports in journals can give an estimate of animal behavior in the wild. Demonstrating cognitive differences in the controlled setting of a laboratory has been much more difficult. Studies of reversal learning perhaps come closest to showing learning differences among species. In these tests, choice A is first rewarded (usually by food) whereas B is not. When the animal has learned to prefer A, the choice is reversed, and B is now rewarded and A is not.

Primatologist Duane Rumbaugh did a series of reversal tests of this sort with a variety of primate species.[36] He found that great apes (especially chimpanzees) did better than monkeys, which in turn did better than lemurs. There was a high correlation between measures of brain complexity and the ability to transfer learning from one trial to another.

The ability to generalize learning rules from one set of stimuli to another might also be taken as a measure of behavioral flexibility. A series of tests comparing corvids (crows and ravens), which have high brain:body ratios for birds, with pigeons, which have relatively low ratios, indicate differences in learning.[37] Species of about the same size are used from both groups in the same testing apparatus. The apparatus has three keys that can be illuminated from inside with a color or a line pattern. In a test for sameness, the center key might be blue, and if a side key is blue and the bird pecks at it, it is given a food reward, but choosing a different-color side key, such as green, is not rewarded. The trial is then shifted to center-key red, side-key red or yellow, and the test is whether the bird can transfer the rule for sameness to another set of colors. Corvids show evidence of rule transfer, whereas pigeons begin at chance levels.

The learning rule is then made more abstract by shifting from color matching to line matching. Birds trained to pick the same color as the center key are asked to pick the same line pattern (horizontal or vertical) between the center key and a side key. Corvids perform at high levels from the beginning, whereas pigeons continue to perform at chance levels.

Learning experiments have been criticized on a number of grounds. Every animal is likely to have talents specific to its niche. Thus, we might expect rats to be good at mazes because they continually run through narrow passageways, whereas pigeons would do much worse on such tests. There are also differences in sensory specializations. Rats do poorly on tests of visual discrimination but learn very quickly when olfactory cues are involved.[38] There is also the problem of motivation. How can we be sure that both groups of animals being compared are equally hungry, for example, for the food reward? The problems with learning sets are extensive enough that some researchers, such as psychologist Euan Macphail, can claim that no quantitative or qualitative differences in learning have been proved between any vertebrate species.[39]

Flexibility and Play

Despite the problems of defining intelligence, there are clear behavioral differences in large-brained animals. Play is found almost exclusively in warm-blooded species: mammals and birds, which have about 10 times the brain:body ratio of groups that evolved before them. There are few reports of play in amphibians and reptiles. It is most prevalent in birds with the largest relative brain size and in the large-brained primates, and species that continue play into adulthood tend to be the largest-brained members of their taxa.[40]

Play is associated with complex behavior patterns, long learning periods, and extended dependency of the young.[41] A long period of parental care protects the young and apparently affords time to experiment with novel behavior and learn complex skills. In primates, play is positively correlated with degree of immaturity and length of dependency.[42] In birds, the most playful species (e.g., parrots and songbirds), are all altricial, being born relatively helpless and needing extended parental care, whereas play is nearly absent from precocial waterfowl, which are quickly independent of their parents.[43] A statistical study of 26 bird orders found play in significantly more altricial than precocial orders.[44] Carnivorous mammals, especially members of the cat (felid) and dog (canid) families, have long periods in which they learn hunting skills and engage in extensive play.[45] A large study of mammals found a significant correlation between EQ and play when making comparisons between orders of mammals, but this did not remain true for comparisons at the level of families or species. The researchers note that other factors may be at work besides brain size alone, and that they used fairly crude measures of play.[46]

Types of play are often related to the kinds of skills an animal will need as an adult. Prey species like rodents and deer practice running during solitary play, whereas young carnivores tend to prefer object manipulation, especially if it resembles prey items.[47] A study of domestic kittens confirmed what many a pet owner knows: fur-covered toys are much more likely to be pounced on and bitten than furless objects.[48] A bird study of captive American kestrels gave two groups of fledglings either (1) fur-covered mouse mimics, pine cones, and twigs; or (2) corks, pine cones, and twigs.[49] Both groups of young raptors played an equal amount of time with these objects, but only the group with mouse mimics preferred those toys significantly more of the time than the other choices. It is interesting that adult kestrels did not manipulate any of these objects.

Play is associated with species on high-energy diets and high metabolic rates. Species with low-energy diets (e.g., leaf eaters) play less than those on high-caloric diets like frugivores, carnivores, and omnivores, which typically have higher metabolic rates also.[50]

Play is usually reserved for the juvenile period and often seems to serve as practice for adult behaviors. Herring gulls, for example, like to drop sea shells on hard surfaces like roadways to crack them and extract the contents—a kind of tool-using behavior. Yearling gulls play at dropping vegetation or flotsam on soft surfaces like grass.[51] A study of cheetahs in the wild found that cubs that frequently stalked and crouched at family members during play also stalked and crouched at prey more often, and cubs that had more object play and social contact play also had more contact with live prey when it was released near the young by their mother.[52]

Deprivation studies suggest there is something essential about play for proper maturation. Young rats engage in play fighting, and social isolates, raised without companions, become socially incompetent, reacting either too aggressively or defensively in later social encounters, and they are also sexually incompetent.[53] It is interesting that if they are raised with nonplayful partners, they still show these defects, suggesting there is something about the need for play itself, rather than mere companionship, that affects maturation.[54]

Play is associated with flexibility, with trying out a variety of behaviors and roles. Adults will exhibit a more restricted pattern of these behaviors than juveniles. Dominant animals reverse roles and play subordinate in play fighting, especially among canids.[55] Young polecats during play fighting show most of the aggressive motor actions they will use as adults, but also a variety not seen again later.

By extending the juvenile period, a species might gain extra time to master difficult skills of group living.[56] A study of 27 species of primates found that the proportion of time spent as a juvenile was positively correlated with forebrain size and also with average social group size. While juvenile primates can often forage as successfully as their parents, it takes longer to master the complexities of social life, such as coalition forming, cooperation, and reconciliation.[57]

Life History Changes

A large brain is a slow-growing organ, and this fact results in a long period of dependency of the young. This means parents will have to supply food for a long time, and they will tend to have smaller litters to limit their costs. The reproductive years may be extended to compensate, and the whole life cycle will likely be stretched out. Slowly maturing young take longer to reach sexual maturity, and this is another reason that the reproductive years, and life in general, are likely to be extended. This lengthening of the life cycle is seen repeatedly in large-brained primates, birds, toothed whales, and mammals in general.[58] A comprehensive survey of 493 mammalian species found a significant correlation between large relative brain size and longer life spans with longer reproductive periods.[59]

A large, slowly maturing brain has its benefits, however. It retains plasticity and allows for a long period of learning. The overlap of generations allows skills, and perhaps even culture, to be passed from one generation to the next. The high energy needs of a large brain will likely entail a shift to high-energy foods that are generally difficult to secure. Eating grass and leaves does not take a lot of skill, but hunting down an herbivore requires expertise that needs a long learning period. The slow maturation of carnivores allows such periods of apprenticeship. Likewise, extracting nuts from shells, or finding widely disparate fruiting trees, requires complex skills that are learned by other slowly maturing species.

The characteristics described in this chapter are therefore related to each other. A large brain facilitates complex behavior, both in social group organization and foraging skills. The slow maturation of a large brain means a long juvenile period that emphasizes flexibility, innovation, and play. This behavioral flexibility may even be an advantage in new environments or in periods of rapid ecological change that threaten extinction to less flexible species. This is part of the "cognitive buffer hypothesis" discussed in chapter 2 that suggests that mammals and birds survived the extinction of the dinosaurs because they were able to quickly adopt new feeding techniques.[60]

Primates have been large-brained, long-lived mammals almost from their beginnings, and they exemplify the advantages of stretching out the life cycle. Primates have the longest juvenile periods relative to body size of all mammals. Sequential delays in the major growth stages of life permitted a longer period for brain development as well as for other parts of the body. For example, human leg bones, which are crucial for bipedal walking, are much more developed than those of apes because of our longer growth period. As Sue Taylor Parker and Michael Mc-Kinney point out, an extended period of fetal growth, when all cortical brain cells originate, leads to a bigger brain as humans mature.[61] Moreover, an extension of infant and juvenile stages allows more connections (synapses) between brain cells to develop. As neurons mature, they are insulated (myelinated). Myelination in rhesus monkeys continues until three and a half years of age, but humans are unique in extending this process through the teenage years and into early adulthood.[62]

Extensions of the life cycle are associated with the level of parental investment in a variety of species. A mouse lemur can begin reproducing when it is one year old and produce two litters of up to three offspring per year. A gorilla will not reproduce until it is about 10 years old and has a single young about every four or five years.[63] These differences are associated with the r and K life history strategies discussed in chapter 1. Extensive parental investment is necessary in species that rely heavily on learning for survival. It favors fewer young and an extended juvenile period as well as an extended life cycle in general. A slowly maturing brain facilitates behavioral flexibility and learning, especially in regard to foraging skills and social relationships.

In chapter 3, I showed that humans build their big brains by extending fetal rates of neuron division for about two years after birth, and that this may have been accomplished by changes in a small number of genes regulating growth. Other singular human characteristics can be attributed to an extension of juvenile growth patterns. Figure 4.1a shows a general primate characteristic: the head is a much larger proportion of the body in youth than in adulthood. By retaining juvenile growth patterns, humans maintain a high brain:body ratio. Figure 4.1b shows how similar the skulls of chimpanzees and humans are at birth. But as adults, the chimp has diverged much more from the juvenile pattern than has the human.

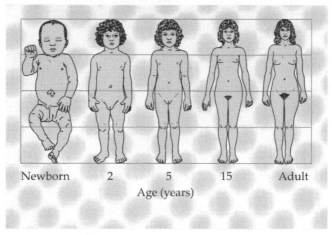

(a) Differential growth rates in a human

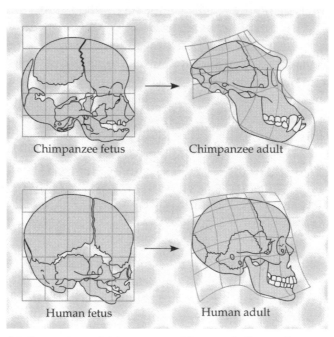

(b) Comparison of chimpanzee and human skull growth

Figure 4.1 Proportional growth of the skull.
(*a*) The human head occupies a much larger proportion of body size in youth than as an adult. (*b*) Grids show that the proportions of the human skull are very similar to those of the chimpanzee at birth, but in adulthood the coordinates have changed much more in the chimpanzee than in the human. (From Neil A. Campbell and Jane B. Reece, *Biology*, 6th ed. [San Francisco: Benjamin Cummings, 2002], 478. Adapted by permission of the publisher)

Our bodies also remain hairless, a general characteristic of primate newborns. We can go on learning our whole lives, remaining playful and innovative as adults to a much larger degree than other mammals. We are also hypersexual, a general characteristic of youth. Most animals have a mating season and periods of "heat." Humans can be randy year-round.

Neuroscientist Lori Marino suggested there may be "universal" consequences to having a large brain: social complexity, behavioral versatility, learning, and innovativeness.[64] Humans excel in all these qualities, and we have seen them emerging in a variety of lineages. We have also had unusually rapid rates of evolution and been able to invade diverse habitats worldwide.

Humans are not just an accidental species that appeared on one branch of the primate lineage. Our constellation of qualities has been tried repeatedly in nature, and we are only the latest and most extreme example. K-selected strategies that build behavioral versatility into slowly maturing offspring are some of the fundamental ways life uses to remain adjusted to the challenges of an ever-shifting environment. It may not be too far-fetched to say that a species like us is in the cards wherever evolution has had long enough to play out its possibilities. In the next two chapters, we will explore four different groups of animals that share these "humanlike" characteristics and see that the high-information pathway is a strategy that has been evolving in a variety of lineages.

Life began by storing information in genes and thus being able to pass body plans from one generation to the next. It then evolved nervous systems that allowed faster responses to environmental change than genes alone could afford. A few animals have evolved culture, such as hunting techniques or even ways of using tools, that are passed by observational learning from one generation to the next. Humans extended the process of information gathering with the invention of complex language, then printing, computers, and the Internet. In the last chapters of the book I will explore the idea that increases in complexity and information content may in fact be a cosmic process that applies to both the organic and inorganic realms.

5

A Constellation of Qualities

In this chapter and the next, I will sketch portraits of four species that represent peaks of brain development in their lineages: chimpanzees, ravens, elephants, and dolphins. All have high brain:body ratios and share qualities of behavioral complexity and versatility. I will argue that a large central nervous system is the source of a wide repertoire of behavior. It is unusually flexible and innovative, enables complex social relationships and rich communication, and appears capable of a degree of abstraction. The association of these qualities in widely differing lineages and different environments suggests they are somehow connected with each other in the evolutionary process. The gathering information content of life has expressed itself in similar ways in different lineages where the main common anatomical characteristic is high relative brain size.

Members of these species are highly individual and appear to value each other as such. There are many individual differences in behavior, and this may be reflected in cultural differences between small groups of animals. They have much to say to each other in terms of complexity of communication, and they manipulate their environment in complex ways. Dexterity comes from widely different appendages: a hand, a beak, or a trunk, depending on the species, but in each case there is an unusual intricacy of operation. A large nervous system appears to create a self that is seeking to express itself, both within the species and in reworking its surroundings.

These animals appear to value each other to an unusual degree, coming to each other's aid even when they are nonrelatives, and holding attitudes toward death that in some cases look like mourning. Alliances and greeting ceremonies indicate intense relationships between individuals. Again, there appears to be a self that is emerging at a higher level, where it can view itself, others, and its surroundings with a new objectivity that allows individual recognition and the ability to see regularities in nature, form concepts, and combine them in new ways.

The extra neurons of a large brain not only lead to more complex behavior, they may supply the ability to abstract patterns from experience, to form concepts. Mimicry is present in all of the species we will consider, and it could be argued that you have to have a sense of the other before you can imitate its actions. There appears to be a mental representation of a model that the animal is trying to match. We know that a mechanism like this takes place in song learning in birds:[1] a young songbird forms in its memory a template of a singing adult, which serves as its model before the maturing bird tries to match it with its own vocalizations. Thus behavioral models can be set in memory for some species of animals.

Tool use in these species also seems to require a certain abstraction of qualities, such as an understanding of the hardness of stone or wood that is used to break something softer or more brittle. I will argue in a later chapter that the extra neurons of a large brain could be used to monitor patterns in earlier-acting neurons and thereby abstract features from raw material. This is, in a way, a forest and trees problem. You have to be a certain distance above the forest to see patterns within the forest. Some neurons could be used to monitor patterns of response in other neurons, and in this way abstract features from the primary data. We will see repeatedly "third-order" relationships in the behavior of these species. An animal not only responds to another animal, it can form strategies in relation to a third animal. They not only make tools, they may make a tool to modify another tool.[2] Concepts appear to be at work, as when an animal carries a tool a long distance and it is just the right size or hardness for the food it will extract.

These species are extreme cases of the K-selected, equilibrium species that use a wide repertoire of behaviors to adjust to a changing environment. At the end of this chapter and the next, I will elaborate on their evolutionary significance. These four species come from different lineages and live in different environments: two on land, one in the air, and one in the sea. What they share in common is not ecology or genetics, but a high brain:body ratio.

CHIMPANZEES

Brain Size

The chimpanzee is our closest living relative, and we shared a common ancestor about 5 to 6 million years ago. Chimpanzees have an encephalization quotient (EQ) of about 2.4, which means their brain size is about 2.4 times greater than an average mammal's of their body size. The EQ of humans is about 7.5, some three times larger. Humans, chimpanzees, and most of the other great apes have a special class of neurons in the forebrain known as *Von Economo neurons* (VENs). In both humans and chimpanzees they appear late in gestation.[3] The size of these cells is strongly correlated with relative brain size in primates, whereas the size of neighboring brain cells is not.[4] They are most numerous in humans and are also present in two other species, elephants and dolphins. Their large size is believed to facilitate the rapid exchange of information between frontal brain areas, and they are implicated in both self-awareness and social cognition.

This section will focus mainly on the behavior of the chimpanzee (*Pan troglodytes*) but will also have some comments on a sister species, the bonobo (*Pan paniscus*), which separated from chimpanzees about 2.5 million years ago.

Social Complexity

Chimpanzees in the wild live in fission-fusion societies of 30 to 100 individuals in which large groups may split into smaller groups for a while and rejoin later. On a given day, an adult male may be part of a large hunting band and then go off foraging on his own, and individuals may stay apart for days before rejoining their troop.[5] In spite of this freedom of association, there are clear dominance hierarchies in the larger group, one for males and another for females. Researchers can easily determine these by the submissive gestures made by subordinates to higher-ranking individuals. Each group has an alpha male who won his position through contests with other males, and he has first rights to food and estrous females (during their fertile period).

Many animals live in dominance hierarchies, but there are special qualities in chimpanzee society that hint at more complex relationships. In the first place, alpha males cannot simply be despots. They often owe their position to coalitions, and two lower-ranking males can put a stronger single male to flight. One of the partners may then become the alpha male, but he in turn will defend his partner in disputes with lower-ranking individuals. These alliances may last nearly a decade, and although they are often between kin, they also occur between unrelated individuals.[6] Moreover, the rest of the troop, including the females, can have an important effect on the tenure of a ruler. In one case where a young, inexperienced male became head of a colony, he often favored high-ranking males and females. As a result, other females formed coalitions against him and kept him away from settling disputes. Within months he lost his alpha position, and leadership went to an older, more even-handed male. Thus, although there are clear hierarchies in chimpanzee society, governing depends to a certain extent, on the consent of the governed.[7] Primatologist Frans de Waal, who has studied primate societies extensively both in captivity and the wild, says an enduring leader of a group of chimpanzees is often more interested in keeping the peace than in favoring relatives or allies. In a large fight, he will separate combatants, literally prying locked fingers apart and giving a blow to anyone who continues fighting. In this sense he places himself "*above* the conflicting parties."[8] That is, he seems to have an abstract sense of what is best for the troop, above loyalty to any individual.

Chimpanzees can also have complex relationships during a hunt. In some areas of Africa, meat is an important part of their diet. In the Taï forest of West Africa, the preferred quarry is the colobus monkey, a midsized animal that can climb higher in the forest canopy than the larger, heavier chimps. Males of a troop form coordinated hunts where some chimps climb into trees to flush out the monkeys, others drive them in particular directions, still others block particular escape routes, and finally an ambusher, who has to anticipate where the fleeing monkey will wind up, is stationed to grab it. The sophistication of this coordinated attack is shown by the fact that it takes a chimpanzee some 20 years to become a good ambusher, for he must be able to calculate not only the activities of other members of his troop but also the reactions of the panic-stricken prey. This has been described as an understanding of tertiary relationships, for a chimpanzee must understand his relationship to both other hunters and the quarry.[9] The coalitions in leadership struggles can also be thought of as tertiary relationships, for an individual must keep in mind both his relationship with his partner and their combined effect on other members of the troop.

Primatologists Michael Tomasello and Josep Call[10] cite eight reasons to believe that primates understand third-party social relationships, among them the facts that individuals (1) "redirect their aggression preferentially toward kin of their past enemy," (2) "simultaneously threaten a dominant individual while appeasing another individual dominant to the first," (3) "respect one individual's 'ownership' of other individuals and objects," and (4) "encourage reconciliation between third parties who have just been fighting." Although many animals live in dominance hierarchies, the complexity of relationships in chimpanzee society has rarely been found elsewhere.

Complexity of Communication

Chimpanzees are capable of a lot of individuality in their communication with each other. One study distinguishes 32 different calls used to express 13 different emotional states.[11] Moreover, these calls can be graded in intensity through changes in pitch, loudness, and duration. Intensity is also conveyed by the changing nature of the call; increasing distress, for example, is conveyed by whimpers, then squeaks, and finally screams. There are also dialect differences between different communities across Africa, reflected in the pitch and duration of calls.[12] Individuals that join a new troop have been found to modify their calls over time to sound more like those around them, much as a human who moves to the South might gradually acquire a southern accent. Chimpanzees can also imitate each other quite precisely.

The chimpanzee face, like that of humans, has many sets of small muscles that allow a rich repertoire of facial expressions to emphasize vocal communication. They also have a variety of gestures that can be used flexibly, and there are differences between populations that again suggest culture.[13] For example, stripping leaves with the mouth makes a distinctive sound that gains attention. Of 30 individuals in an East African group, eight use it in a sexual context, whereas five use it in play. In the Taï forest of West Africa, males use it as part of a drumming display. One study recorded eight different gestures to initiate play. Body language can also be used deceptively.[14] One chimp used a submissive gesture after a fight, and when its opponent approached, it attacked again.

Experiments with artificial languages suggest that chimpanzees are capable of higher-order abstractions in their understanding of vocalizations. The throat of a chimpanzee is not bent at the right angle to produce the variety of sounds that humans can make, so their sound repertoire is limited. But they have successfully been taught Ameslan, American Sign Language, and they have also mastered keyboards with abstract symbols to represent a variety of nouns and verbs. Experiments with these kinds of communication indicate that chimpanzees are capable of concept generalization and even understand syntax.

Lana is a captive-born chimpanzee that learned to use a keyboard with lexigram symbols for a variety of objects and actions.[15] She was able to generalize terms she learned to new objects. For example, she became fond of drinking Coca-Cola, and when offered a Fanta orange drink for the first time, she called it "the Coke which is orange." She had been given an apple and a symbol to represent it, but after her first encounter with an orange, she asked for the "apple which is orange." In a similar way, her first cucumber was labeled the "banana which is green."[16]

Kanzi is a captive-born bonobo, which is a sister species of chimpanzees.[17] He observed his mother being taught a keyboard of lexigrams and he was not directly rewarded, as she was, for successful responses. Kanzi learned to respond to hundreds of lexigram symbols as well as to spoken English. He could comprehend complex sentences such as "Take the telephone outdoors" or "Give Rose a shoe." In multiple tests of this kind, Kanzi responded correctly 59 percent of the time, whereas a two-year-old human child scored 54 percent. He also combined words with gestures in novel ways, such as pointing to the person who should perform an action, or pointing at the location toward which an action should be directed. Perhaps most remarkably, he understood syntax so that "Pour the juice in the egg" meant something entirely different from "Pour the egg in the juice." Kanzi performed correctly 81 percent of the time (71 out of 88 instances) in requests of this nature.[18]

Syntax is a higher-order structure than words alone. It requires an understanding of the relationship between words. There is no inherent reason why one word is the subject and another word the direct object unless you understand their order in the sentence. Understanding syntax appears to require a degree of abstraction that occurs in other places in chimpanzee cognition, such as in deception and tool use. The calls of chimpanzees in the wild are not known to have this complexity or syntactic structure. This suggests that some species in the wild may have latent abilities that do not appear until the right cultural or ecological circumstances invoke them. We will see this theme again in the behavior of other large-brained species.

Deception

Chimpanzees appear to be capable of "theory of mind," taking into account both what they and another animal are thinking. They are not simply reacting to the context of a situation, but evaluating what other players know before deciding on a strategy.[19] This is a difficult topic to define precisely, because how do we really know what is going on in the mind of another individual? When a business associate sits across the table from you and smiles agreeably, can you really be sure of what he is thinking? How much more difficult must this be with an animal that does not have language as complex as our own?

Deception is found all over nature. Many butterflies, for example, have bright colors on the inside of their wings and cryptic colors on the outside. If a predator approaches, they quickly close their wings and blend seamlessly with their surroundings. If undisturbed, they open their wings, revealing bright colors that can attract a mate. Does the insect *intend* to deceive when it closes its wings? And what of the amazing wood grain on the outside of its wings that makes it blend so well with tree bark? Surely the insect did not intentionally paint itself.

Evolution explains the mechanism like this: brown butterflies that were more woodlike on the outside had an advantage—they were less picked off by predators like birds, so that over time variants with better camouflage became more numerous in the population. The process is a bit like what Michelangelo said when asked how he carved his statue of David. He said that starting with a block of marble, he chipped away everything that was not David, until there it was. Evolution is often negative sculpture of this sort, selecting against less fit variants until the survivors are very well adapted to their niche. The colors on butterfly wings are under genetic control, so genes can produce a deceptive pattern. Can this ever be distinguished from a mental decision to deceive? I think most people would agree that humans are capable of flexible plans that mislead others. Can any other animal do this?

Observations from the wild and also controlled experiments indicate that chimpanzees can intentionally deceive. In the wild, an alpha male will try to monopolize all the estrous females, but there are usually too many randy males and females around to keep track of all of them. Sometimes a lower-ranking male will induce an estrous female to join him behind a rock or large bush, out of sight of the dominant male, for some clandestine lovemaking. They will move carefully and quietly, glancing back at the alpha male to make sure they do not alert him.[20] Besides choosing a concealed place, they appear anxious about what the alpha male knows.

De Waal describes a chimpanzee in his colony that hurt a limb while fighting with a younger male.[21] For a week, he limped whenever he walked in front of his rival but returned to a normal gait as soon as he was out of sight. Because injuries inhibit aggression, de Waal believes he was trying to deceive the younger male.

In a carefully controlled lab experiment, food was made available to a chimpanzee on either side of a booth.[22] Inside the booth sat a human who could act as a competitor by taking away the food. If the human stared at food on one side, the chimpanzee preferred to grab food from the other side. If the human oriented his body to side A, but turned his head to watch side B, the chimp preferred to snatch food from side A, where the eyes of his competitor were not staring.

In the next set of trials, the chimpanzee could come around the side of the booth and take the food (the direct approach), or he could first walk behind an opaque barrier and come up to the side of the booth and then take the food (the indirect approach). The opaque barrier was set up under one of two conditions: (1) it was placed so that the human competitor in the booth would see the chimpanzee first walk *away* from the food, and then the primate could sneak up unobserved and snatch it; or (2) it was placed so the human would not first see him walking away.

The critical result is that the chimpanzee took the longer, indirect route only when the human would first see him walking away from the food. He appeared to want his competitor to see his deception (appearing to ignore the food), and if his rival could not see it, the primate took the shorter route. Other tests show that chimps are concerned with what others see.[23] They follow the gaze of other chimpanzees, will move to get a better view of what the other is looking at, and glance back at them if nothing is there. All of this looks very much as though they are taking account of what others know and flexibly forming plans based on that knowledge—although it will always be difficult to be certain what is going on in the mind of another individual, animal or human.

Chimpanzees also appear to take account of what rivals hear. A human was placed in a booth that had two clear tunnels with food at the end, which a chimpanzee could reach from the outside. The human could also take the food away. The door to one tunnel was noisy, and the other was not. When the human was looking to the side, so he could not see what the chimpanzee was doing, the ape preferred to take food through the silent door. When the human was not present, there was no preference in doorway.[24]

Tool Use

The complexity of chimpanzee social relationships is mirrored in the complexity of their material culture, and here also a degree of abstraction is evident. Chimps are proficient tool users, and some of their manufacturing is eerily reminiscent of early human techniques. In the Taï forest of West Africa, 26 different kinds of tool use have been catalogued, including implements for honey dipping, termite fishing, cleaning wounds, sponging liquids, aimed throwing of missiles, and pounding nuts.[25] Nut cracking, like hunting, requires a long period of apprenticeship, and it involves some of the most complex skills.[26] Five different nuts are eaten in the Taï forest, and during the four months when they are plentiful, chimpanzees get most of their calories from them. Chimps have to transport hammers over long distances to sites where the nuts are plentiful, and they have to anticipate the hardness of the shells they will encounter. For soft nuts, wooden hammers suffice, but for the hardest ones, stones must be used. The fruiting tree is beyond the line of sight, so this suggests a calculation must be made for the tool they take along.

Some nuts require precise techniques. For example, hard Panda nuts need a powerful blow to smash the shell, but then gentle taps to extract the kernels. The nut has to be repositioned two to three times while being worked on. Youngsters can begin to pound nuts in the first two years of life, but it takes a year of practice to make positive gains from soft nuts, and seven years to work the hardest nuts usefully.

Figure 5.1 shows the most sophisticated kind of anvil for nut cracking as found in Bossou, Guinea. On top is the hammer used to crack nuts against the stone anvil below it, but most astounding is the bottom stone that is used to keep the anvil level. Primatologist Tetsuro Matsuzawa calls this a *metatool*, for it is a tool used to modify another tool.[27] We are not far here from *Homo habilis*, a human ancestor who left behind stone choppers and scrapers dating to about 1.8 million years ago.[28]

Figure 5.1 Tools made by chimpanzees in the wild.
A stone anvil (*middle*) supported by a wedge (*bottom*), with a hammer (*top*) used to crack nuts, assembled by wild chimpanzees in Bossou, Guinea. (From T. Matsuzawa, "Chimpanzee Intelligence in Nature and Captivity: Isomorphism of Symbol Use and Tool Use," in *Great Ape Societies*, ed. W. C. McGrew, Linda F. Marchant, and Toshisada Nishida [Cambridge: Cambridge University Press, 1996], 201. Adapted by permission of the publisher and T. Matsuzawa)

The hominid tools were most likely made by a flaking technique in which stone chips were knocked from a larger stone by hitting it with another stone. Nothing this sophisticated has been observed among chimpanzees, although occasionally a stone hammer will break at Bossou, and the smaller fragment gets used as a superior hammer. But it is not hard to visualize a continuity from what these chimpanzees are doing to the tools of early human ancestors.

Using tools in sequence might also qualify as metatool use because the activities of the first tool are needed before the second tool becomes effective. A report from Gambia describes four tools used in sequence to extract honey from a hive. A female chimpanzee was observed breaking off a dead tree branch to produce a stout tool with a sharp end, which she used as a chisel to break through hardened material the bees had made to protect the side of their nest in an old tree stump. She then used a shorter branch to widen the hole she had made with the first. She then trimmed a green branch and jabbed it into the nest area, but when it did not produce honey, she abandoned it and made a fourth tool. This was longer and made from a green vine, and it enabled her to extract large amounts of honey. These tools were used in progression, the first two as chisels and the second two for dipping into honey. Each had a different size and was held in a different manner by the animal, the first two in a power grip to chisel, and the second two like fishing probes to extract the honey.[29]

Serial tool use was also observed among wild chimpanzees in the Congo among groups that add termites to their diet. Two wooden tools may be used in sequence to extract the insects from their underground nests.[30] First a stout stick is used to puncture the ground over the nest and then a thinner stick is used as a fishing probe to pick up the insects. The fishing probe has a brush tip that typically takes three operations to manufacture: after breaking a stem from a plant, the leaves are removed and a brush is created by sliding one end of the stem between partially closed teeth. Like good painters, the primates wet the brush with saliva and then compact the fibers by hand before using it. The puncturing sticks have a characteristic length and 98 percent are made from a single species of tree. The fishing probes also have a characteristic length but are half as thick, and 96 percent are made from a different plant.[31] Chimpanzee communities in East and West Africa also go termite fishing, but different groups use different tool kits.

I discussed earlier the degree of abstraction that was implied by understanding syntax in the language-trained primate Kanzi. There was no inherent reason for understanding subject and direct object unless one understood their order in a sentence.[32] In a similar way, there is no inherent reason why a stick and a termite nest should go together unless one has a sense of the qualities of each—that is, the hardness of the wood needed and how it might be used to break into a termite tunnel. It appears to be an abstraction of the nature of each and a comparison of their qualities. Matsuzawa makes a very similar argument for "hierarchical cognition" in chimpanzees in third-order tool use, as when they use a wedge to prop up an anvil that is the surface for a hammer.[33] We have also seen third-order social relationships where a coalition of chimpanzees work against a third individual, and where a hunter has to keep track of both chasers and quarry in deciding where to position himself to ambush prey. In chapter 7, I will discuss brain mechanisms that may explain this ability to form hierarchical relationships and their role in the process of creating abstractions.

Cultural Differences

Cultural differences abound in wild chimpanzees and are probably a result of individual differences in learning, although genetic factors cannot be entirely ruled out.[34] A survey of 42 different types of tool use in six different chimp populations across Africa found a different tool kit in each of them.[35] Taï chimps share only 57 percent of their tool types with chimps in Gombe, that in turn share only 54 percent of their tool kit with those in Mahale. Techniques also differ: in Gombe, chimps insert twigs into termite nests to fish for them, whereas in Cameroon they first break open the termite mound to gain direct access to the insects. In the Taï forest, two types of nut that are eaten by troops on one side of a river are not eaten by chimpanzees that live on the other side, even though the same trees grow on both sides. Local culture rather than ecology is the most likely reason for the difference.

Individual discovery and cultural diffusion are probably major reasons for local cultures. New techniques are discovered in different places and can spread through a troop by social learning. An amusing example of tool discovery is illustrated by Mike, a relatively small adult male in Jane Goodall's research area in Gombe, East Africa.[36] Mike, perhaps because of his size, never fought with the other adult males, the usual way to ascend to alpha status. One day, however, he discovered empty kerosene cans left by the researchers, and banging them loudly in front of him, he charged at a group of dominant males. They fled in terror, and he repeated this performance until they crept back with submissive gestures and began to groom him, a sure sign of newly acquired status. He was able to maintain a dominant position in his troop for about six years.

Mike's insight was not adopted by the other chimpanzees, but there are cases where new discoveries spread through a community. Adult males like to produce long-distance sounds by drumming on the buttresses of trees. At Mahale, researchers built metal houses that the primates simply walked past for several years.[37] But beginning in 1979, some males began drumming on the metal walls, and this quickly became a habit of most male passersby. There was a lot of individuality in how this was done. Some drummed with both hands, some with one, and others used both hands and feet. There were also differences in loudness and rhythm.

Researchers were able to witness another example of social learning when they supplied unfamiliar Coula nuts to a troop in Bossou.[38] Although adults took little interest at first, the more exploratory juveniles tried to crack the nuts. Adults gradually took up the habit, and the numbers making use of this new food went from 11 percent in 1993 to 67 percent in 2002.

Individual differences of culture have been found in various chimpanzee populations across Africa, not only in terms of tool use but also in hunting techniques, courtship, grooming, play, and a "rain dance" (a slow display at the start of rain). A survey of nine study sites across the breadth of Africa found 39 cultural variants where a behavior common in one community was not present in another.[39] The most striking differences were in skills that require long periods of learning: tool making, hunting, and forming social coalitions.[40]

The insights into noise making and using a new food resemble examples of insight learning found in lab experiments with chimpanzees. A famous one is psychologist Wolfgang Kohler's experiment in the early part of the last century, in which he suspended a banana from the ceiling with a string out of reach of chimpanzees and supplied boxes and a stick in another part of the room.[41] The primates figured out a way to stack the boxes, climb on top, and use the stick to knock down the food. These experiments have been replicated in more recent times also. The solution to this problem was found faster than mere trial-and-error learning would allow and suggests that disparate elements were put together in a conceptual way.

The Value of the Individual

Chimpanzees go to great lengths to maintain individual relationships, and grooming is an important ritual that helps create social bonds. It also serves the basic physical need of removing parasites and dirt from the skin. Alliance partners spend much time grooming each other, and the larger the social group, the more time spent in grooming, reflecting the greater complexity of relationships.[42] When chimpanzees meet after a long separation, they often fling their arms around each other with cries of excitement and then indulge in a session of mutual grooming.[43]

After conflicts, which are a natural part of any animal society, there are multiple ways to reestablish peace. A subordinate may approach a dominant individual with signs of submission such as bowing, hand gestures, and cries of distress. If the dominant wishes to reconcile, he will embrace, pat, and kiss the other, and they may then settle down for a period of mutual grooming. After a fight, a subordinate seeking reassurance often reaches out his hand in a supplicative way that is very recognizable from human behavior, and if the dominant accepts the proffered hand with his own, reconciliation usually follows.[44]

Goodall relates an incident at Gombe in which a male attacked an older female severely enough to wound her.[45] She later approached him with submissive gestures and screaming, and he embraced and patted her until she quieted down. He then groomed her for 10 minutes, and she became so relaxed that she fell asleep at his feet. Goodall suggests that embracing and kissing, which are a regular part of chimpanzee reconciliations, probably evolved from maternal behavior in reassuring a child, and a dominant and subordinate are seeking to recreate this atmosphere of trust. Kissing is probably derived from the mouth-to-mouth transfer of masticated food from a chimpanzee mother to her child. Bonobos even French kiss, and this is probably derived from the mother's use of her tongue to transfer food.[46] The human versions of these unusual ways to express affection in animal societies probably have similar origins.

The importance of the individual in chimpanzee society is also shown by the care given to members that have been hurt. Large wounds can occur from territorial fights or leopard attacks in the jungle. Care is extended to all members of the group and is not restricted to relatives. In evolutionary theory, altruism is supposed to take place mainly between relatives who share each other's genes to various degrees. Altruism between nonrelatives has rarely been found in nature, although it has been observed in the care of injured chimpanzees. Aid to wounded chimps can be elaborate and includes cleaning wounds with saliva (which has antibiotic compounds), removing dirt, and chasing away flies.[47] A female wounded by a leopard was tended for the rest of the day and her wounds licked for more than four hours.

When a female chimpanzee was killed by a leopard in the Taï forest, the cat was chased off by the dominant male. Subsequently, six males and six females were seen sitting silently near the body. Three of the males groomed the body for more than an hour, although they had never been seen grooming her before. In all, the body was attended by some members of the troop constantly for six hours and 15 minutes after her death. When a two-year-old youngster fell from a tree and died, his mother picked him up and gave loud alarm calls for 10 minutes. She was guarded by the dominant male for the next two hours and 50 minutes, and could still be seen the next day carrying the body close to her chest, although by now it was swollen and beginning to smell.[48]

Altruism is evident in tests done with semifree-ranging chimpanzees in Uganda. These animals are fed by humans at night, but the experiments were done with a person with whom they had never interacted before. In the first test, a human is put behind bars and reaches for a stick that is beyond his grasp. Twelve of 18 chimpanzees on the other side of the bars handed the stick to the human, although they did not earn any reward for their actions. In a second similar test, the apes had to climb over an array of obstacles to help the human, but a significant number still did it. In a third experiment, a chimpanzee is put in a cage with a door that leads to a second cage with food in it. The door is held shut with a chain attached to a peg in a third room that has another chimpanzee in it. The second animal has the option of releasing the peg so the first chimp can open the door and get the food, even though the liberator will not share in the food. A significant number of chimpanzees release the peg. As a control, they have a second peg they can release that controls a door that leads to a room without food.[49] Another study of wild chimpanzees in Uganda found that the majority of cooperative activities among males was between individuals who were not genetically related to each other. The researchers evaluated six social behaviors, including grooming, coalition forming, meat sharing, and joint patrolling of territory boundaries.[50]

The unusual degree of cooperation among group members in chimpanzee society is perhaps mirrored in the fierceness of hostility that can be expressed to out-group members. Wild chimpanzees defend territories, and males regularly patrol the borders of these areas. If a lone male encounters a gang of neighbor males at their mutual border, he may be viciously attacked. Chimpanzees are among a small group of mammals that kill their own kind, another way in which they are eerily reminiscent of our own species. Goodall witnessed the annihilation of a small community by a nearby larger group.[51] One by one, males of the smaller group were ambushed by the larger neighboring group. In some cases, while one male held the victim down, others viciously bit and pounded him, so that he usually died soon after of his wounds. When the last of the males and older females had been killed, the larger troop took over their territory and added the remaining young females to their group.

Perhaps most shocking is the fact that both these groups had once been tied to each other by bonds of kinship and friendliness. They came from one original group that split when it reached a size of about 32 adults. One group took over the southern part of its range and had its own dominant male, while the other, larger group retained the northern part of the territory and had a different male leader. The occupation of different territories signaled a complete change in social relations.

Goodall says the attacks between rival groups were much more violent than the encounters usually seen between males competing for dominance within a single social group. They were more like chimpanzee attacks on large prey like monkeys. They were treating rival neighbors like prey.[52] Perhaps the intensity with which they see "not-self" is part of the intensity with which they view the self. In-group members are treated with an empathy that is rare in animal societies, their wounds attended to and their anxieties soothed by reconciliation. Out-group members are treated with rare hostility. Although many animal species defend territories and squabble and display at the borders of their areas, few reach the deadly hostility of rival chimpanzee communities. This warlike behavior has been also documented at other chimpanzee study sites, in Tanzania and in the Taï forest of the Ivory Coast.[53]

This all, of course, has echoes of some of the worst aspects of human behavior. In both cases, it may come from a combination of the ability to form concepts and the normal aggressive behavior between rivals. Many animals fight over territory and mates, but rarely to the point of death, since the goal is only to decide the stronger one, and inflicting great harm is counterproductive to both winner and loser. But if these contests are heightened with concepts, so that the opponent is seen as little more than prey, extreme violence may be triggered. Dehumanizing the enemy is always the first step before genocide in human warfare. The enemy must be seen as less than human. Perhaps, as Goodall suggests, the killer chimps see out-group males as less than chimpanzee. The mental powers that let them abstract concepts in tool making may allow them to demean the enemy to the point where he is no more than a thing to be disposed of.

Self-Recognition in a Mirror

If chimpanzees view self with a special intensity, it is perhaps significant that they are one of a small number of species that have been shown clearly to recognize themselves in a mirror. Most species, on viewing their image in a mirror, treat it as a stranger. Many people have had the experience of a bird that pecks at the outside of a reflective window, outraged at the apparent rival who seems to imitate all his motions. A few large-brained species—such as chimpanzees, bonobos, orang-utans, elephants, magpies, and dolphins—act in a way that indicates self-recognition in front of a mirror. The mark test, as it is called, usu-ally involves sedating an animal and putting brightly colored marks in places they cannot normally see, such as over an eyebrow or ear, or just under the chin. The dye is odorless and nonirritating. After waking up, the animals are supplied with a mirror. Chimpanzees can be seen touching the marked parts of their bodies, and smelling their fingers afterward, much more than they do with other parts of the body. No monkey's movements in front of a mirror are as clearly self-directed as those of chimpanzees.[54] Chimps do other things in front of a mirror that indicate self-recognition, like picking food from their teeth, or grooming parts of the body they could not otherwise see.[55]

There are endless controversies about these tests, and some biolo-gists believe they prove nothing significant.[56] But in the context of the discussion here, they are part of a mosaic that fits a general pattern. Chimpanzees have high relative brain size and complex societies and vocalizations, and they are capable of innovation and local culture. A survey of 116 different primate species found a positive correlation between forebrain size and frequency of innovation (inventing a new behavior), social learning, and tool use.[57] At one level, we may say that a large brain facilitates more complex behavior. But I suggest that it also creates a greater sense of self that has more to say to conspecifics (richer communication), values them more highly (helping behavior and elabo-rate reconciliations), and manipulates its world in intricate ways (tool use). It is as though there were an emergent self that views itself and the world around it with a new objectivity. Recognizing the image in a mir-ror as self rather than as a stranger may be an indicator of this increased self-awareness. These are characteristics of the high-information path-way that relies on flexibility of behavior and third-order concepts as strategies for survival.

It may not be surprising to find such a human-sounding constellation of qualities in our nearest living relatives, with whom we share about 99 percent of our DNA. But in fact we find a similar set of qualities in very different lineages where the main thing we share in common is a high brain:body ratio.

RAVENS

Brain Size

Corvids include crows, ravens, rooks, jays, and magpies, comprising about 120 species in all. They belong to the songbirds (oscines), which are part of the largest and most recently diversified order of birds, the Passeriformes. Songbirds generally have high brain:body ratios, and corvids have the largest brains of all the songbirds.[58] The EQ of a crow is very similar to that of a chimpanzee.[59] Metrics that rely on absolute brain size would lead one to expect very different cognitive abilities in corvids and apes. The raven's brain at 14 grams is much smaller than a chimpanzee's at 390 grams, yet there are many convergences of behavior. In chapter 3, we reviewed the evolution of brain:body size in mammals and birds and saw that it had increased about tenfold over that of reptiles, amphibians, and fish (see figure 3.4). The area that has grown the most in mass in both mammals and birds is the forebrain (the telencephalon), which is involved in "executive functions" like learning, memory, and social interactions. The brain stem has changed much less between species and is mainly involved in various automatic responses like controlling blood pressure and breathing rate. Another way to estimate brain evolution is to compare the size of the forebrain and the brain stem.[60] A study of 140 bird species found this ratio highest in corvids and parrots. It is interesting that comparing the forebrain:brain stem ratio in corvids to that of other birds yields results that are similar to comparing the ratio for primates to that of a group of primitive mammals, the insectivores.[61] The forebrain in birds is built very differently from that in mammals; it does not have the six-layered structure of the one in mammals, yet it appears to be the seat of executive functions in both groups.[62] The pathways for vocal learning form similar loops in both mammalian and avian brains.[63]

Large forebrain size in birds is positively correlated with social complexity, vocal learning, feeding innovations, and long periods of parental care.[64] Ravens have especially long periods of continued parental care after the young have fledged and left the nest.[65] This section will focus mainly on ravens but will also include some of the remarkable abilities of other corvid species, which show many qualities in common with large-brained primates.

Social Complexity

The common raven (*Corvus corax*) is the world's largest crow. It can live for more than 50 years, forms long-term pair bonds, and may mate for life. The main social unit is the family, but ravens also associate in larger numbers in fission-fusion societies where they cooperate temporarily for specific tasks.[66] Ravens apparently know each other individually, because they will arrange themselves in strict dominance hierarchies when feeding. The dominant bird will perch at the top of a carcass, and 10 to 15 birds will align themselves on either side of the meal in order of dominance. The lowest-ranking bird may actually be getting just as much food as the alpha male, but he will be farthest from the head of the table.[67]

Ravens are able to form alliances with each other and with other species to get what they want. There are numerous reports of groups of them harassing large animals like wolves and eagles in order to steal their food. One raven will come up behind an animal and pull its tail with its beak, while another raven, with precise timing, flies off with the distracted animal's meal.[68]

Ravens apparently recognize humans individually by multiple cues.[69] Bernd Heinrich, a biology professor at the University of Vermont, has spent decades studying ravens in the wild as well as in aviaries built next to his home that allow him to observe their behavior daily. Four of his hand-reared ravens allow only him to approach them. If anyone else comes closer than 15 yards, they fly up in fright. When they were two years old, he did some experiments to find out how they recognize him. When he went into the aviary with a stocking cap pulled over his face terrorist-style, they were not alarmed. Wearing clothes they had not seen before also did not faze them, but when he came in wearing a bear suit, they acted frightened. Next he had a female neighbor wear his clothes, with a stocking cap pulled over her face, and the birds flew around the cage in alarm. Heinrich concluded that the birds must know individual humans by multiple cues. A report from Cornell University relates that a biologist who regularly invades crow nests to band young fledglings was singled out for attack by adult birds as he crossed campus in the midst of a crowd of other faculty and students.[70]

Ravens appear to be capable of forming alliances with other species. They are found so often with wolf packs that Heinrich subtitled one of his books *Investigations and Adventures with Wolf-Birds*.[71] Without the sharp bills or talons of a raptor, they must rely on other animals to make the kill and cut open tough hides. Data on 24 wolf kills in Yellowstone National Park found an average of 32 ravens at a carcass.[72] Stories from Eskimos, as well as other native peoples, also suggest a long-term association between human hunters and ravens.[73] On the tundra or ice, a flying bird will have a much wider view than an earthbound animal, and it would be in the bird's interest to lead hunters to promising game where they might share in the spoils. Some Eskimo hunters believe ravens even indicate to them the location of prospective prey by dipping their wings during flight. If this seems far-fetched, Heinrich asks us to consider the honeyguide (*Indicator indicator*), a small African bird that leads the honey badger (*Mellivora capensis*) to hives that the badger opens, and then both species share in the sweet rewards. In some areas of Africa, the bird has even transferred this behavior to humans.[74]

Rooks, a highly social corvid species, are apparently sensitive to third-party relationships. They will redirect aggression against the partner of an individual that has been aggressive toward them.[75] They also appear to engage in a kind of consolation behavior. After a conflict, both victim and aggressor engage in postconflict affiliations with their respective partners, as evidenced by bill twinning, in which they interlock their beaks with their social partner.[76]

Deception

In deciding how to act, ravens and other corvids apparently take into account what other birds know. Ravens and jays like to cache food by burying it in various locations. Ravens are quite versatile at this; after digging a hole with their beaks and placing the food inside, they will cover it with whatever is nearby: dirt, grass, leaves, sticks, or snow.[77] Given an excess of appetizing items, they will cache first and eat later because they know how greedy their own kind can be. Ravens will readily steal from another bird's cache given a chance. A study of wild ravens, as well as ravens in an aviary, shows that caching ravens will interrupt their burying of food if they see they are being observed by another raven.[78] If observed, they are likely to move behind a large obstacle, such as a rock or tree, to start a new hole, or they dig up food and move it to a new place when the original site was observed. Prospective thieves, in turn, will wait until the cacher departs before trying to raid the site.

Carefully controlled experiments are needed to show that ravens have "theory of mind"—that is, really take account of what others are thinking. An experiment put two cages in a room with an uncaged raven that was caching food.[79] One cage contained a raven with a clear window on the front, and the other cage held a raven behind an opaque window. After the uncaged bird had buried its food, a second set of trials was run five minutes later in which the caching bird was put back into the room with either the observing bird or the one behind an opaque window. Results show that cachers recovered significantly more buried food when put together with an observer that had been behind a clear window than with a nonobserver. They seemed to take account of what the other bird knew and recovered their food when its location had a greater chance of being known to another.

These results are similar to those of experiments done with chimpanzees in which it is shown that food retrieval varies depending on what a competitor knows. Recall that chimpanzees also appear to want a competitor to see their deceptive practices, and that they take shorter routes to food when they are not observed.[80]

A similar experiment was done with another corvid, the Florida scrub-jay (*Aphelocoma coerulescens*), which also indulges in caching and pilfering.[81] A bird was given a large amount of a favorite food to cache away in the presence of another jay in a cage that had either a clear or an opaque panel on the front. After three hours, the cacher was allowed back into the room to recover his stash in private. Because there had been an excess of food, the bird ate some and reburied the rest. The crucial result is that if the bird had been observed by another bird, he moved his remaining caches to new locations significantly more times than if he had been with a jay that could not see him behind an opaque panel. He not only remembered the locations of his stash, but the social context in which he had acted earlier. This provides strong support for the idea that the birds take into account and remember what others know.

It is interesting that the prior experience of these hand-raised jays seemed to influence their attitude to thieving.[82] If the caching bird had itself pilfered other birds' sites earlier, then it reburied its stash, as already described. But if the bird had no prior experience of being a thief, it recached its food significantly less often, even though it knew it had been observed by another bird. This supports the old adage that "it takes a thief to catch a thief."

Vocal Complexity

One study of raven vocalizations recorded 81 different call types, many of them specific to individual pairs of ravens.[83] Although some calls are probably innate, such as the loud *caw* sound that is given by both crows and ravens during states of excitement, other calls are probably learned and can be specific to a small number of birds.[84] There are local dialects; Heinrich reports different calls in every area he has visited outside of New England.[85] Ravens are also known to sing for hours, combining a wide variety of sounds in a kind of musical warble that is quite different from the harsh *caw* that is familiar to most people.[86]

Besides voice, body posture is also used to express mood.[87] Head feathers have their own muscle control, and Heinrich says he can readily see from its body language whether a raven is afraid, self-assertive, angry, or contented.

Ravens also have a remarkable capacity to mimic, and when missing a mate will call out with the partner's own individually specific call.[88] In addition, they are able to mimic a wide variety of nonavian sounds, including overheard human activities. Wild ravens have been reported to make the sounds of radio static and a motorcycle revving up. One naturalist in Olympic National Park heard a raven say, "Three, two, one, bcccchhhh" (the latter being an explosive sound). Park rangers had been conducting avalanche control the previous week and doing countdowns before setting off dynamite. Elsewhere in the park, ravens would sometimes sit atop urinals that had automatic flushing. Off in the woods they could be heard making the gurgling sound of water rushing down a drain.[89]

Behavioral Versatility

Ravens may be the world's most behaviorally flexible birds in terms of diet and distribution.[90] They are found around the globe from Arctic tundra to high mountains, deserts, forests, and urban environments. They feed on meat, insects, fruits, and grains. Besides scavenging large kills of other animals, as described earlier, they themselves hunt small game, such as smaller birds, by knocking them down in midflight. They dig up bird nests from the ground and can fish by wading into streams and grabbing fish with their bills.[91] Young ravens are extremely curious about novel items, although they become more cautious as they get older. During a 1991 Easter egg hunt in Juneau, Alaska, 1,200 colored eggs were hidden. While people were registering kids for the event, ravens came and made off with most of the prizes. The local paper's headline was, "Don't count your eggs before the ravens come."[92]

Different species of corvids have some remarkable specializations, probably facilitated by their large brain size. Clark's nutcracker (*Nucifraga columbiana*) is the avian champion of memory and has been recorded to hide away up to 33,000 pine seeds in 6,600 separate sites over a wide area and find them as long as six months later.[93] Another corvid, the western scrub-jay (*Aphelocoma californica*), remembers where it has buried food as well as how perishable it is. In one experiment, birds were allowed to bury perishable food (the larvae of wax moths, a preferred food item) and more stable food (peanuts).[94] When released again after four hours, 80 percent of the birds went to the caches with larvae. But if they were released after 124 hours (when the buried larvae would have decayed), they all went first to the peanut caches. They therefore have both spatial memory and clock memory for food items.

Tool Use

A survey of avian tool use reported in 68 journals from around the world found 125 cases in 104 species. The most numerous examples were among the corvids, with the common crow having the most techniques, including using stone hammers, sharpening pieces of wood to use as probes, and descaling fish by scraping them over sand.[95]

Perhaps the most remarkable tool maker is another corvid that makes a variety of tools, the New Caledonian crow (*Corvus moneduloides*).[96] It produces a stick about 13 centimeters (5 inches) long with a hook on one end that it is believed to use to extract food from crevices. Producing the tool is a multistep process that involves breaking off a side twig just above its junction with the main twig. The remainder of this side twig

(a)

(b)

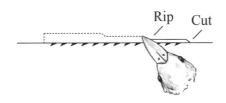

Figure 5.2 Tools made by New Caledonian crows.
(*a*) Hooks on the ends of wood tools. (*b*) (*Top*) Finished leaf tool. (*Bottom*) Cross-cuts and rip cuts made with the beak to produce the completed tool. Part of the saw-toothed edge of the original leaf faces downward. (From [*a*] G. R. Hunt and R. D. Gray, "The Crafting of Hook Tools by Wild New Caledonian Crows," *Proceedings of the Royal Society B—Biological Sciences* 271 [2004]: S89, and [*b*] G. R. Hunt, "Human-Like, Population-Level Specialization in the Manufacture of Pandanus Tools by New Caledonian Crows *Corvus Moneduloides*," *Proceedings of the Royal Society B—Biological Sciences* 267, no. 1441 [2000]: 404, 408. Used and adapted by permission of the publisher)

will form the hook. It then breaks off the main twig just below this junction and strips off any remaining leaves. Finally it uses its bill to sculpt and sharpen the hook (figure 5.2*a*). While a variety of animals are known to use tools, these are the most modifications in shaping a tool by any known bird.

These birds, which live on two tropical islands in the southwest Pacific, also fashion tools from leaves of the Pandanus tree. It takes multiple cuts to produce these implements that are then held in the beak at the wide end and used to probe into tree holes for food.[97] The finished tools may also be stored on perches for later use. Figure 5.2*b* (*top*) shows a finished product; the saw-tooth edge was part of the outer edge of the original leaf from which the tool was cut. Figure 5.2*b* (*bottom*) shows the multiple cross-cuts and rip cuts made by the beak of the bird to produce the finished device. There are also cultural differences in tool production in different parts of the islands. In one area, tools are shorter and have more steps to them because there are more cross-cuts.[98]

The intricacy of these tools recalls the multiple, stepwise modifications of wooden tools made by chimpanzees in the Congo to extract termites from underground nests. Concepts appear to be at work in multiple mechanical relationships, as when chimpanzees use both a wedge and a hammer to extract nuts on an anvil.

It might be argued that tool making could be just an instinctive process, like the web spinning of a spider, which it knows how to do from birth, but there are reasons to think that insight learning is involved with these birds. This has been tested by presenting them with novel problems they would not find in nature, and the speed with which they solve them rules out a simple instinctive solution or trial-and-error learning.

For example, a wild-caught New Caledonian crow was given the problem of extracting a bucket of food placed inside a pipe in an aviary.[99] The bucket had a looped handle at the top. Straight pieces of wire about 9 centimeters (3.5 inches) long were also made available in the cage. The crow bent these wires, some with hooks very similar to the ones on wooden tools found in the wild. Inserting these bent wires into the loop of the bucket, she was able to lift it out and get her meal. Her only previous experience with this kind of material was pipe cleaners that had been supplied to her one hour before the experiment began and which she was allowed to manipulate freely. When subsequently supplied with wires, she successfully bent them and retrieved food nine out of 10 times. Her technique involved holding the wire down with one foot and then bending it at an angle with her beak. It took her about 40 seconds to complete this manipulation, and she was able to retrieve the food within two minutes. This looks very much like insight learning with a new material that she is not likely to have known from the wild.

A similar experiment has been done with another corvid, the rook (*Corvus frugilegus*). Four hand-reared birds were presented with hook tools that they could use to retrieve an out-of-reach bucket holding food. Although they had never seen this tool before, three used it successfully on the first try and the fourth on the second try. Next they were presented with straight wires, and all four successfully bent them into hook shapes that they could use to retrieve the bucket, even though they had no experience of making such tools before. Rooks are not known to use tools in the wild, and this suggests that some species may have latent abilities that can be invoked by the right cultural or ecological experiences.[100] We saw something similar with the ability of Kanzi to use syntax with words.[101]

The New Caledonian crow also appears to be capable of metatool use, as with the chimpanzees that used one tool to modify another. Seven wild-caught crows in an aviary quickly learned to retrieve meat that had been placed inside a 15-centimeter (6-inch) hole with long sticks that experimenters supplied to them.[102] In the next phase of the experiment, the birds were individually tested with two tool boxes. One contained the long stick, which was out of reach of the bird's bill. In the other tool box was a stone in a similar position. In front of the tool boxes was a short stick, too short to retrieve the meat but long enough to reach inside the tool box to get either the long stick or the stone. Six of the seven crows tried to retrieve the long stick with the short stick on the first try, even though they had no previous experience of obtaining one stick with another. Three of the seven completed both tasks on the first try and got the food reward. All were eventually successful, although some took longer to master both challenges. This looks like insight learning for most of the birds because there were not enough tries to learn by trial and error. The researchers consider this metatool use because two tools were used in series to obtain a desired result.

Corvids, as mentioned earlier, have especially high brain:body ratios for birds. An anatomical study shows that the New Caledonian crow has an especially large brain size among the corvids, with the main increase coming in the associative areas of the forebrain.[103]

The common raven is also capable of insight learning. Six hand-reared ravens were presented with meat dangling by a string from a branch on which they sat.[104] The string was 76.2 centimeters (2.5 feet) long, and they had never seen food presented in this way before. Five of the six birds solved this problem in four to seven minutes. They pulled the string up with the beak, put a foot on the loop to hold it in place (much as humans do when tying a bow), and then pulled up the next section of string. This might have to be repeated five to six times to get at the food. One bird did the entire sequence correctly the first time he tried it. To guard against social learning, each of the six ravens was tested separately, out of sight of the others.

I believe this can be called insight learning, as we saw with the chimpanzees that stacked boxes to get at dangling bananas, because there was not enough time to have learned by simple trial and error. It suggests a conceptual understanding of the problem and a quick solution that jumps past alternative possibilities.

Like most birds, ravens build intricate nests, but they also use their bills in a variety of ways to manipulate their surroundings. They dig in snow or dirt to cache food, as already mentioned. One report describes a raven in an outdoor cage that tried to dig its way out from the inside as two other wild ravens dug under the netting from the outside, finally enabling the captive bird to make its escape.[105] Other intricate uses of the beak include prying and breaking wood, gripping and tearing meat, and combing and caressing the feathers of a conspecific.[106]

In chapter 4, I discussed the connection between behavioral flexibility and play. Play affords a chance to practice actions used in later life, and novel behaviors are often observed during play. Large-brained, slowly maturing animals are often noted for long periods of play in youth. Ravens are notorious for play, both as adults and when young. They have been observed rolling down snow-covered roofs of houses, then climbing back up for another ride. Young birds will slide down a snow-covered hill on their breast and belly feathers. Ravens have been seen dropping a variety of objects—such as rocks, sticks, turf, and even sheep droppings—in flight and then swooping down to catch them again before they hit the ground. A person hang gliding reported seeing about 10 ravens play in flight with a 6-meter-long (20-foot) white plastic streamer. One would swoop down with the streamer in its beak and then pass it on to the next raven, which performed similar antics.[107]

In general, Heinrich remarks on the individuality of these birds and their ability to come up with novel behaviors he has never seen before: "[M]any raven individuals are unique and any one person observing them can hope to see only a very small portion of the species' amazing repertoire of behaviors."[108]

In the next chapter, I will discuss the unusual responses of elephants to death of their own kind and suggest that their complex, long-term relationships may make them value each other to an unusual degree. Ravens have unusual reactions to other dead ravens.[109] Although they are fond of the meat of all kinds of birds, they are apparently reluctant to eat another raven. When Heinrich provided his ravens with a dead raven shot by a hunter, they would not touch it. He gave them a dead headless crow, but they would not get near that either. When he skinned a crow, removing the head, wings, and legs, they accepted the meat readily. Ravens are black, and when presented with a whole black chicken carcass, they had no qualms about devouring it. It is not the color, or even the meat, but the resemblance to a dead raven that apparently unnerves them.

Self-Recognition in a Mirror

The magpie (*Pica pica*) is a corvid in the crow family that also has food-storing habits. Working with five hand-reared adult birds, experimenters put brightly colored red or yellow marks on the black throat feathers under the bill, where they could not normally be seen. One experimenter shielded the eyes of the bird while the other affixed self-adhesive colored marks. For controls, they put black marks in the same place, in the same way, on the black throat feathers.[110] When supplied with a mirror in the cage, birds with bright marks showed self-directed behavior, trying to touch the marks with the beak or foot. Birds without a mirror, or with black marks on the throat, showed little or no self-directed activity. Also, birds with bright marks that could see themselves in the mirror stopped the self-directed motions when they had removed the marks. Birds were tested one by one so they could not see each other's reactions. This is the first avian species to show this kind of self-recognition in front of a mirror.

The lineage that led to ravens and the lineage that leads to chimpanzees have been separated for at least 250 million years. The avian brain also has a very different architecture from the brain of mammals; its surface is smooth while the mammalian brain is convoluted with many fissures. In addition, the avian brain lacks the six-layered organization of the surface cortex found in mammals, although in both lineages, large brain size comes from increased growth of the embryonic forebrain.[111] Ravens and chimpanzees live in different habitats and have very different body sizes, but they have in common a large brain:body ratio. They also share many behavioral similarities related to the complexity of social organization, vocal communication, and the ability to mimic, innovate, and manufacture tools. Concepts appear to be at work in the ability to modify tools for a special use and carry them to new places where they will be used. There are local cultures and great variability in individual behavior. A large brain appears to be creating a greater self that is expressing itself in complex relationships with the animate and inanimate worlds. Both groups have a long period of youth and dependency, long lives, and an extended playfulness that may facilitate the inventiveness found in adults. Both taxa appear to recognize themselves in front of a mirror in ways that are rare in the animal world. Ethologists Nathan Emery and Nicola Clayton have suggested there may be universal consequences to having a large brain. It is a flexible tool kit that includes causal reasoning, imagination, and the ability to anticipate future events.[112] In the next chapter, I will explore two more groups that live in very different habitats but share many of these same behavioral qualities.

6

The Evolution of Personality

The qualities considered so far in chimpanzees and ravens—complex societies and communication, innovation, tool use, play, and individual recognition—have an oddly human ring. We too exemplify these traits. This may sound wildly anthropomorphic, but in fact my intention is the opposite: to indicate that this constellation of qualities has been emerging in a variety of lineages, of which *Homo sapiens* is only the latest and most extreme example. They are a suite of characteristics found in large-brained, slowly developing animals that use behavioral versatility and insight learning to survive in their niches. They are an extension of the K-selected, equilibrium strategies that life has used since earliest times as one way of remaining adapted to an ever-changing environment. Life has increased in information content over time, and a summit of this process is not a mind that acts like a faster computer, but a personality with wide behavioral versatility. It is an emergent self that has more to express to conspecifics and in its manipulation of its surroundings. This self has a high level of objectivity regarding social relations and tool use, and this manifests itself in inventiveness, insight learning, and great individuality of behavior. These all sound like human characteristics because we are all on the high-information pathway as a way to deal with the challenges of a fluctuating environment.

In this chapter I will present two more species that live in very different habitats and have very different brain architectures but that follow a similar pattern.

ELEPHANTS

Brain Size

Encephalization quotient (EQ) measurements for elephants vary widely, from about 1.7 to 2.3, but they compare favorably to those of chimpanzees at 2.2 to 2.4.[1] Elephants have the largest brain of any terrestrial animal in absolute terms, weighing 5,000 grams (11 pounds) in the Asian elephant as opposed to about 1,400 grams (3.1 pounds) for a human and 440 grams (slightly under a pound) for a chimpanzee. As brains become larger, the area of the surface, the cortex, increases faster than any other region of the brain. Most higher associative functions have been located in this cerebral cortex. Thus, the elephant, with its especially large brain, should have many "extra" neurons available for higher-order brain functions, despite the fact that its large body size brings its EQ down to more average levels. Neuron density is lower in elephants than in human brains, so that the overall number of cells in the larger elephant brain is probably less than in humans, though precise counts have not yet been made.[2] A special kind of neuron, the Von Economo neuron (VEN), is present in the elephant forebrain. These large cells have previously been found only in humans, great apes, dolphins, and related cetaceans, in similar forebrain areas in all four groups, and they are believed to be involved in complex social behavior.[3] Elephants also have an unusually large hippocampus,[4] an area associated with long-term memory, perhaps giving credence to the legend that "an elephant never forgets."

Elephants are a classic K-selected species with large brains, long lives, delayed reproduction, and a long period of socialization. The long period of dependency of young elephants allows for an extended period of learning and socialization, as it does in humans. Elephants are nutritionally dependent on their mothers until four years of age, socially dependent to ages 10 to 16, and can live more than 60 years.[5] The African species first gives birth at about 14 years, with intervals of about four to five years between further offspring.

Three species of elephants are now recognized: the African and Asian varieties and a forest elephant of which little is yet known. The Asian and African species are estimated to have diverged from each other about 5 million years ago.[6] The modern Asian species, *Elephas maximus*, is believed to have evolved from earlier elephant species after the mid-Pleistocene, about a million years ago.[7]

Social Complexity

African and Asian elephants live in complex, multitiered societies whose core is the family: a mother, and her subadult offspring. Individual families are often joined into larger "family units" comprising several mothers and their young and led by a matriarch, the oldest and most experienced female in the group. Males have little to do with this social organization; in their midteens they leave the family unit and lead solitary lives or associate in loose bachelor herds. Their subsequent contact with females is during periods of male heat, or musth, when they seek out receptive females. Daughters, however, tend to remain with their mothers and raise their own young within their natal group, leading to the strong matrilineal structure of elephant society.

When family units get too large, a subgroup breaks off, so that nearby social groups may retain a high degree of relatedness.[8] Family units associate in larger "bond groups," and these in turn join in larger aggregations called *clans*, which may number up to several hundred individuals.[9]

Elephants have a fission-fusion pattern of association, where smaller groups join into larger groups for a while and then separate at other times, such as during the dry season when resources become scarce.[10] The core of this organization is a mother and her offspring, but almost as enduring is the family unit, consisting of up to 20 mothers with their young.[11] One study shows a high degree of relatedness among these mothers, who are sisters from either the current or previous generation. Analysis of the larger bond groups also shows relatedness at the aunt-niece level or closer. At the clan level, elephants are not necessarily closely related, and one study suggests they form clans for social reasons and to attract mates more than anything else.[12]

Cynthia Moss, who has spent many years studying wild elephants in Africa, wrote in 1988 that all the 650 elephants in the Amboseli National Park knew each other.[13] An adult female may be familiar with the calls of up to 100 other adults. This has been described as the largest network of vocal recognition of any mammal.[14]

The importance of knowledge in elephant society is indicated by the fact that a group is usually led by an older, more experienced member. Females 35 years and older usually lead larger social groups than do younger matriarchs.[15] Playback experiments give an indication of some of the advantages of an experienced leader.[16] When hearing recordings of different elephant calls, older matriarchs were better able to distinguish familiar from unfamiliar calls and to lead their group in the best defensive maneuvers. The age of the matriarch also predicted the number of calves per female in her group, indicating that older, more experienced matriarchs provide direct reproductive advantages to their group.[17] Reproductive advantage is the standard for fitness in evolutionary theory. It means more of a particular gene will be passed on to the next generation. Because the females in a group of elephants are usually related, this means that a wise old matriarch that protects her group has a greater chance of having her family's genes become more widespread in the gene pool of the next generation.

Complexity of Communication

Elephants have a complex vocal repertoire that includes rumbles, growls, snorts, barks, roars, and trumpets.[18] Most are produced by the vocal cords, but some originate in the trunk. At least 30 different types of calls have been catalogued according to their social context, including contact calls, distress calls, alarms, and sexual signaling.[19] The widest variety of calls are made by females, reflecting both their large brain:body ratio and more complex social activities. The calls of African and Asian species appear to be similar. Elephants may also be capable of vocal imitation. One study reports an adolescent female living in a camp in Kenya that appeared to imitate the sound of nearby trucks.[20] Spectrum analysis showed sounds that do not resemble ordinary elephant calls but that have frequency characteristics very similar to the truck's. In another case, an African elephant that spent 18 years in a zoo with two Asian elephants made chirping sounds characteristic of Asian elephants that are not usually found in its own species.[21] Statistical analysis showed these chirps were close to the sound qualities of its Asian "mentors." Although vocal tutoring is common among songbirds, it is rare in mammals.

Elephants can also make infrasounds that travel long distances.[22] The human ear hears frequencies between about 20 to 20,000 hertz (Hz), but elephants can produce sounds as low as 5 to 24 Hz that propagate over great distances. In the evening on the African savanna, temperature inversions facilitate these signals, creating listening areas as great as 300 square kilometers (115.8 square miles). Elephants make most of their low-frequency sounds during this period, and more than 27 different low-frequency rumbles have been identified.[23] We are still a long way from understanding the complexity of this communication, but as mentioned earlier, an adult female may be familiar with the calls of as many as 100 other individuals, an unusually large network of vocal recognition for any mammal and an indication of the subtle variations the elephants' vocalizations probably contain.[24]

In addition, elephants have complex chemical communication, mainly from hormones secreted by glands. An elephant's nose may be five times more sensitive to smells than a bloodhound's, an indication of the importance of this channel of communication.[25] Social excitement is almost always accompanied by secretions that dribble down the sides of the face from temporal glands located on either side of the head. These secretions are a complex mixture of compounds and may signal a variety of conditions, such as sexual receptiveness in both males and females, distress, pregnancy, or illness. In addition, volatile compounds have been found in the breath and urine of elephants that elicit different behaviors in the receiver, ranging from aggression to attraction, depending on the chemical secreted.[26]

Eighty different visual and tactile displays have also been described in elephants. Combined with chemical and sound communication, these offer a wide repertoire of messages that can be conveyed between individuals.[27]

Tool Use

The dexterity of an elephant's trunk has been compared to that of the human hand. The trunk even has structures similar to our opposable thumb. The end of an African elephant's trunk has two extensions, whereas an Asian elephant's has one. These can be folded over to grasp a variety of objects by a pinching method similar to what we do with two fingers (figure 6.1).[28]

Elephants can hold objects as thin as a straw or carry more than 272 kilograms (up to 600 pounds) between their trunk and tusks. Using the trunk alone, they are able to pick up a peanut, crack it open, blow away the shell, and pop the kernel into the mouth. There are no bones in the trunk to give it support, and it is operated by a rich network of muscles and nerves. The tip has especially dense nerve endings and small hairs, giving it great tactile sensitivity. Dissection of an Asian elephant's trunk has found some 150,000 muscle subunits along its length.[29]

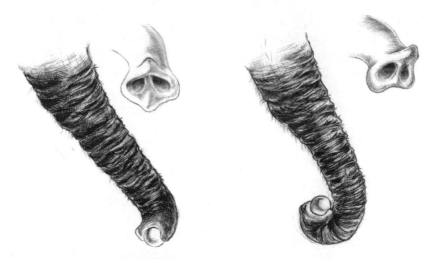

Figure 6.1 The dexterity of elephant trunks.
Extensions on the ends of the trunks of the African elephant (*left*) and the Asian elephant (*right*), showing how they are able to grasp an object. (From J. H. Shoshani, "It's a Nose! It's a Hand! It's an Elephant Trunk!" *Natural History* 106, no. 10 [1997]: 38. Used by permission of the illustrator, Utako Kikutani)

Aside from primates, elephants are the most widely cited mammal to engage in tool use. Darwin noted tool use in elephants as an example of animal intelligence in *The Descent of Man*.[30] Wild elephants in India have been observed using the branches of at least five different plants to brush away flies. Like chimpanzees and corvids, they are able to modify objects. This ability is considered a more advanced form of tool use than the simple utilization of objects as they are found in nature. Elephants have been observed holding down a branch with a foot and then twisting off side branches or the end to make it a suitable size. They are able to throw objects quite accurately with their trunks to drive off jackals and leopards, and have been known to treat bothersome humans in a similar way.[31] In captivity, elephants have been observed using sticks to pull food closer, open faucets, throw tires onto branches to bring them closer, and make a pile of tires to stand on and reach out-of-the-way branches.[32] It takes about a month for a newborn elephant to learn to use this complex appendage correctly, and until then it may stumble over its trunk in a comical way.[33]

The trunk has a unique evolutionary origin. It is a fusion of the nose and upper lip, with nostrils running in two parallel tubes down its length.

It gives elephants an acute sense of smell for the hormones and other chemical messages they send each other. Besides its great dexterity, it is used socially when two elephants greet each other by entwining trunks in a kind of elephant handshake, and it can also serve as a siphon, sucking up to 37.8 liters (10 gallons) of water per minute into the mouth.[34]

The versatility of the trunk rivals that of the human hand, despite very different evolutionary origins. What they have in common is being part of a high-EQ animal that is expressing itself through intricate manipulation of its environment.

The Value of the Individual

An indication of how elephants value each other is the intensity of their greeting ceremonies. The meeting of two families in a bond group is accompanied by deep rumbling, trumpeting, screaming, and loud ear flapping.[35] The trunk may be used in a kind of elephant handshake, as already noted, and also to examine the face of another.[36] Greetings often involve all sensory modalities: besides touching and vocalizing, excited elephants secrete hormones from temporal glands on the sides of the face as well as urinate and defecate in excitement.[37]

Elephants will go to extraordinary lengths to help a stricken member of their group. Cynthia Moss describes how two family members got on either side of an African elephant that had been shot and tried to hold her up.[38] When she fell over nevertheless, they worked their tusks under her back and head and lifted her to a sitting position. As the elephant was expiring, her family tried frantically to revive her, including placing a trunkful of grass into her mouth.

Elephants appear to have a unique sense of death and mourning. It is, of course, hazardous to infer mental states in another species, but elephant actions are so striking as to appear almost unmistakable. In the previous incident, after the wounded elephant died, family members stood around her, gently touching her with trunks and feet. Some sprinkled loose dirt on her body. Others broke branches from bushes and used them to cover the body almost entirely. They stood vigil throughout the night, and only at dawn began to move on. Such "mourning behavior" is almost unprecedented among nonhuman species. The only other species that are known to show concern for ailing or dead members are also high-EQ animals: chimpanzees and dolphins.[39]

Ethologist Katharine Payne describes the reactions of 129 visitors to the body of an African elephant calf that died near an often-used trail.[40] One hundred twenty-eight changed their behavior in some significant way, including touching the body with the trunk, lifting the body with a foot or trunk, trumpeting, rumbling, and showing guarding behavior. A statistical test revealed that elephants that visited the body were not more closely associated with the calf's mother than those that did not visit; that is, visitors were not necessarily relatives, and males as well as females reacted.

Elephants appear to recognize the skeletons of their own species while being unresponsive to the remains of any other animal.[41] A traveling group will stop and investigate old elephant bones they have not seen before. They seem particularly interested in the head and probe it carefully with their trunks. Although there have been many anecdotes about this kind of behavior, its existence was finally tested for with a controlled experiment. Researchers put arrays of three objects in front of different families of wild African elephants and recorded the results.[42] In a choice between the skulls of an elephant, buffalo, and rhino, they showed significantly more interest in the elephant skull. In a choice between an elephant skull, a piece of ivory, and a piece of wood, there was a clear preference for examining the ivory over the skull or wood. Three of the families had recently each lost their matriarchal leader. They were tested with the skulls of these leaders in different orders, all washed clean to prevent odors that might enable recognition. Each group did not show a preference for their own leader, indicating that elephants cannot tell individuals from bone structure alone (something that would be very difficult for humans also).

Elephant reactions to the bones of their own species are so predictable that filmmakers can lure wild elephants to a spot by leaving old elephant bones there. A playback experiment of the calls of a female that had died elicited contact calls and approach to the loudspeaker by family members 23 months after her death. (This striking behavior is one of the reasons that the Masai, who live near elephants in Africa, believe they are the only wild animal to have a soul.)[43]

Individual Variation

Species with large brain size and long periods of learning may show a lot of individual variation in behavior. There is more to be molded, and a longer time to shape it to different outcomes. Many field observers have remarked on the great individual differences in elephant behavior.[44] This can be seen in the differing responses to the sight of a dead elephant. There is also great variation in the age and speed with which males leave their natal group. Some leave their mothers before age 13, others after 15. Some make the break slowly, coming back for repeated visits, whereas older males, when they finally depart, tend to go for good. They will meet their families again only on the rare occasions when their day ranges overlap. The age of first reproduction in African females can vary from about 100 to 200 months old.[45]

Ages four to 13 are relatively carefree for an elephant, when it is protected by its natal group and it is involved in learning, play, and forming social bonds. Much of the learning in the early years involves food items.[46] The African elephant feeds on at least 91 different plant species, and many of the social contacts of an elephant in the first five years of life involve sampling what others are eating, including placing its trunk into the mouths of older neighbors while they dine.[47]

The importance of the long period of elephant socialization is illustrated by the ways it can go wrong. In areas of Africa that have experienced high levels of poaching that decimate family groups, teenage elephants show behaviors that resemble posttraumatic stress disorders.[48] Females show poor mothering skills, including the rejection of their infants. Fights between males account for 90 percent of male death, whereas normally this happens in only 6 percent of cases in unstressed communities.

The sensitivity to errors in upbringing reflects the long period of socialization of these animals. This, in turn, is a consequence of a large, slowly maturing brain. A complex nervous system fosters complexity of social relations, communication, and intricate manipulation of the environment, as we saw with chimpanzees and ravens in the last chapter. Of the four animal groups reviewed here, only elephants are herbivores. They consume very-low-calorie foliage, so they must spend a large fraction of the day grazing. Yet they show striking convergences of behavior with the other three species reviewed here that are wholly or mainly carnivorous.

Self-Recognition in a Mirror

Elephants are also on the short list of species that show evidence of self-recognition in front of a mirror. Three Asian elephants at the Bronx Zoo were tested individually in front of a full-length mirror.[49] All three did self-inspection in front of the mirror, such as putting the trunk into its mouth or using the trunk to pull an ear forward. These were behaviors not seen at other times. One elephant passed the mark test. A visible mark was put above her right eye, where it would not normally be seen by the animal, and an invisible sham mark, with the same texture and odor as the visible one, was put over the left eye. She repeatedly touched the real mark, but not the sham one, when standing in front of a mirror, and she did not touch either mark before going to the mirror. Although this is a low level of success, it is similar to the percent of chimpanzees that show mirror self-recognition.[50]

DOLPHINS

Brain Size

The bottlenose dolphin (*Tursiops truncatus*) has an EQ of 3.3, higher than any of the primates except humans (EQ 7.5). A number of cetaceans (dolphins, whales, and porpoises) have higher EQs than nonhuman primates, with the highest values found among the dolphins.[51] Neuron densities, however, may be lower in cetacean brains, so counts of total neurons in the brain may be lower than in some ape species.[52] Pod size (size of schools) is correlated with EQ in dolphins.[53] Dolphins have a different brain architecture from that of the animals we have reviewed so far, although we will see many behavioral similarities. For example, primate brains have enlarged frontal lobes, whereas cetaceans have enlarged side lobes, with their frontal lobes remaining relatively small.[54] Most mammals have a six-layered cerebral cortex, whereas cetacean adults lack layer IV, and the uppermost layer I is unusually thick.[55] The folding pattern of the cortex is also very different, and vision and sound are processed in entirely different areas in the two taxa. The unusual forebrain cells called *VENs*—found in humans, chimpanzees, and elephants—are also present in dolphins and other cetaceans. These large neurons are believed to link forebrain areas together and are involved in social functioning and self-awareness in humans.[56] Bottlenosed dolphins have long lives and long periods of dependency, like the other species discussed so far. Females live into their 50s and mature sexually at about eight years old, and males live into their 40s and mature at 10 to 13 years of age.[57] The mammalian lineage leading to dolphins separated from the lineage leading to primates at least 92 million years ago.[58]

Social Complexity

Biologist Richard Conner and colleagues studied wild bottlenose dolphins in Shark Bay, Western Australia, and found evidence of a social network involving more than 400 individuals. Typically, each dolphin had 60 to 70 associates.[59] Adult males form strong associations of two or three individuals that may last for a dozen years or more. These individuals can often be seen swimming in synchrony and cooperating in hunting, defense, and the herding of females. Two of these groups can join against a third group, forming an alliance of alliances, which is rare in the animal world. There are "superalliances" of up to 14 males, with shifting loyalties but clear preferences among alliance partners. Dolphins live in fission-fusion societies in which small groups cooperate with each other for a while and then split up again into smaller associations. Chimpanzees have a similar social organization, but Conner and colleagues claim that bottlenose dolphins have a more complex society with a "hierarchically nested structure."[60] Although there appears to be a dominance hierarchy in groups of dolphins, it is not a simple "pecking order." Lower-ranking individuals may initiate changes in travel direction or be the initiators of chases and social interactions.[61]

The reason that fission-fusion societies appear in all four animal groups addressed in this chapter and in chapter 5 may be that large brain size fosters both greater individuality and the ability to form complex social structures. Certainly humans have the ability to work separately as individuals and also find their place in large organizations.

Dolphin hunting often involves divisions of labor.[62] Some dolphins circle a school of fish; others may hunt by compressing the fish into a ball as some dive into the ball to catch a meal, with apparent turn-taking in these different roles. Bottlenose dolphins have also been observed trapping schools of fish in a bay, with some members acting as guards at the exit of the bay while others moved in to feed.[63] Dolphins will also cooperate in rushing as a phalanx at a school of fish, driving them onto shore and then beaching themselves to feast on the fish flapping in the air over the sand.[64] They are thus using the shore as a kind of tool and feeding in two media: the sea and on land. In the southeastern United States, bottlenose dolphins have been observed creating a bow wave as a group that lifts and strands many fish onto an exposed bank.[65]

Grooming is an important bonding behavior in chimpanzee society. Dolphins also have a kind of grooming in which they touch and stroke each other with their pectoral fins. The dexterity of dolphins is also seen during sex. Foreplay may include stroking with pectoral fins or flukes, back-and-forth rubbing of the pectoral fins, rubbing of the genitalia, and inserting the rostrum into the genital opening.[66]

Tool Use

At a study site in Shark Bay, Western Australia, some of the female dolphins use wild-caught sponges in a tool-like manner.[67] They bring them to deep water channels, holding them near the front of their jaws, and run them over the sandy bottom to flush out fish. When a fish is exposed, they drop the sponge in order to eat the prey, then pick up the sponge again to repeat the process.

In keeping with the great behavioral variety found in dolphin hunting, only about 11 percent of the females use this technique, and the procedure is passed to their young by social learning. Not all calves take up the habit, but those that do are likely to continue practicing it for a decade or more.[68] It is thus a kind of culture or tradition, practiced by a select group of animals and passed on through generations.

Biologists distinguish between observational learning and true teaching. In observational learning, the young pick up a technique just by watching what adults are doing, but in true teaching, the adult must make some additional effort to help the young. How can you distinguish between these two kinds of learning? If the adult's behavior is costly in either time or effort beyond its normal activities, it indicates true teaching. The learning of foraging among Atlantic spotted dolphins fits these criteria.[69] Observations of nine mothers showed that they chased prey longer and made more body-orienting movements to the prey when their calves were present than when they hunted alone. They also seemed to toy with the prey more, let it swim away, bury itself in sand, and then dig it up again. They let attentive calves join the chase and eat the prey in many cases. All these behaviors are costly to the adult in terms of time and energy, and so they suggest that teaching is involved. Tool use has been found in only 0.01 percent of nonprimate mammals, and true teaching is also very rare.[70]

Some food may also be prepared in a tool-like way. Bottlenose dolphins in South Australia like to eat giant cuttlefish. After capturing and killing them, they beat the cuttlefish with their snout repeatedly to release its unpalatable ink. They then turn the body upside down and use the sand on the bottom like sandpaper to scrape way the skin. This releases the indigestible cuttlebone, which floats away, and they then gulp down the remainder whole.[71]

Thirteen different foraging techniques are known among bottlenose dolphins, and at least three of them are passed on by social learning.[72] Techniques include "fish whacking"—that is, striking fish with their flukes and then capturing the stunned prey. They also create walls of bubbles to herd fish and use natural barriers in a tool-like manner. Besides driving fish onto shore, a phalanx of dolphins may swim beneath a school to herd them toward the surface, where they cannot escape. Wild dolphins have also learned to make use of human artifacts. In Florida, they remove baitfish from crab traps, and in several areas they are reported to drive fish toward stationary fishermen's nets as a way to improve their own take.

Imitation

Few species are able to do both vocal and physical imitation, but dolphins are capable of both. Wild dolphin pairs synchronize their swimming behavior, and this synchronization depends on a kind of imitation. In captivity, they can be trained to synchronize a variety of movements. Trainers also report bottlenose dolphins imitating the swimming and grooming behaviors of other species, from fur seals to penguins.[73] They seem to understand human movements in an analogical, symbolic way. If a trainer on land lies on his back and raises his legs, for example, a dolphin in the water may float on its back and raise its tail. If the person waves his arm, the dolphin responds by waving its pectoral fin.[74] In aquaria, dolphins pick up all sorts of behaviors on their own by observational learning. A dolphin was seen trying to remove algae from plate glass with a seagull feather in apparent imitation of a human diver who regularly cleaned the aquarium window.[75] Another was observed using a piece of broken tile to scrape seaweed from the tank bottom, in apparent imitation of a diver who regularly cleaned that area. One report describes a young dolphin that was being viewed by tourists through an underwater window at an aquarium. A man who was smoking emitted a large cloud of smoke. The dolphin calf stared at him intently, swam back to its mother and suckled, and then returned to the window, squirting a large cloud of milk toward the smoker.[76]

Dolphins in captivity have been trained to imitate computer-generated sounds, and they can imitate new sounds accurately on the first try. In one study, when the computer sound was outside the dolphin's range, she could transpose the sound either an octave higher or an octave lower while still maintaining the sound contour of the original signal.[77] To the chagrin of their handlers, bottlenose dolphins will sometimes imitate the training whistles humans use to guide them.[78] Many bird species are capable of vocal imitation, but dolphins (and possibly elephants) are the only mammals known to do this (chimpanzees can imitate physical movements, but not sounds).[79]

There is clear vocal recognition of individuals among dolphins. Each dolphin develops a distinctive signature whistle in the first few months of life; it is learned and distinct from its parents. Wild dolphins use whistle matching, emitting the same whistle type as another dolphin up to roughly 580 meters (1,902 feet) away, apparently as a way of addressing distinct individuals.[80] Biologist Peter Tyack describes a group of six dolphins, of which one female captured by researchers had been placed on a raft.[81] She began imitating the signature whistle of the oldest female in her group, Granny. Granny responded by synchronizing her whistle with the dolphin confined to the raft, and they continued to call back and forth for a half hour, until the captured dolphin was released. No other dolphin in the group participated in this calling, not even the captured female's calf that was nearby.

Whistles are used in many contexts, and depend on learning; each dolphin knows not only its mother's call, but the calls of other members of its group. In an experiment, synthetic whistles were created that removed all normal voice features except frequency modulation.[82] Dolphins could still distinguish relatives from nonrelatives in these truncated calls.

Whistles appear to be used to maintain contact, socialize, and coordinate group activities. Whistling is often heard when a group changes its direction of travel.[83] There also appear to be regional dialects of sound among different groups of dolphins.[84]

Dolphins make sounds in an entirely different way from humans or birds. There is no voice box in the throat with vibrating tissue to produce sound. They use a system of air sacs in the forehead in a way that is still not understood by researchers, but it does not rely on the movement of mouthparts.[85]

Vocal Complexity

Like chimpanzees, dolphins can learn to respond to complex sentences and even appear to understand syntax. Ethologist Louis Herman was able to train bottlenose dolphins to correctly respond to sentences that included both modifiers and verbs with up to five terms.[86] For example, "Surface pipe fetch bottom hoop" meant "Take the surface pipe to the bottom hoop." One dolphin, Phoenix, was trained underwater with computer-generated sounds to guard against unintended cuing by the inflections of a human voice. Another dolphin, Ake, was trained with sign language, using the arm and hand gestures of a trainer. Each method had its own definite syntax. In sign language, for example, the verb usually came last and the indirect object first. Thus, "Right hoop left pipe fetch" in sign language meant "Take the pipe on the left to the hoop on the right." Both dolphins performed correctly far above chance levels on tests with three-, four-, and five-word sentences. What is most remarkable, however, is that they seemed to understand the syntactic nature of sentences. "Hoop fetch pipe," for example, meant take the hoop to the pipe, whereas "pipe fetch hoop" meant take the pipe to the hoop. Both dolphins responded correctly to a majority of reversed sentences of this type, even when they consisted of novel combinations they had not heard before. They were also able to respond correctly to changed modifiers (a "surface basket" was different from a "bottom basket" that lay at the bottom of the pool).

One dolphin appeared to have an abstract understanding of nine different body parts. She could be signaled to touch a basket with either her fin, her tail, or her mouth, for example. Alternatively, she could be asked to shake, display, or touch any one of these body parts. Herman suggests that she has a mental image of her body that lets her use parts in this versatile way.[87] This image may be a higher abstraction of the sense of self than most animals have, and it may explain why high-EQ animals uniquely respond to images of themselves in mirrors.

One of the dolphins seemed to have a concept of missing objects.[88] She was able to indicate correctly whether an object was or was not present in her tank by touching a left- or right-hand paddle when interrogated with sign language. She was 80 to 83 percent accurate on 182 queries of this kind. Also, nonsense instructions, such as the request to take a dolphin to the water, were simply not responded to.[89]

The dolphins could also easily generalize terms to larger classes of objects.[90] The term *hoop*, for example, was taught with a large octagonal plastic hoop, but was immediately understood when the trainers used hoops that were square, round, larger, smaller, lighter, or darker. The experimenters were careful to control for unconscious cuing by the trainers in all these tests.

Dolphins also appear to use their body parts in a symbolic way. They can use their rostrum and body alignment to point an object out to a human swimmer, and they monitor whether the person is attending to them.[91]

Ronald J. Schusterman and Robert C. Gisiner were able to obtain some of these results in language training of sea lions (*Zalophus californianus*) with methods similar to Herman's.[92] Although sea lions have less than half the EQ of dolphins, they still have above-average brain size for a mammal of their body size.[93] Schusterman claims that the feats of both dolphins and sea lions can be explained by just two learning rules:

1. If an object is designated by a signal, perform the designated action on that object.
2. If two objects are designated, and the action is "Fetch," then take the second object to the first.[94]

However, these two rules have many of the qualities of a language: alternative objects, modifiers, and verbs can be placed in the "sentences" in the proper places, and the animals understand their relationships in novel combinations. Arbitrary sounds or gestures represent objects or actions. A certain order of symbols stands for a certain order of activities in the real world. These, I submit, are some of the basic hallmarks of language comprehension of any kind. Work with sea lions shows that this level of symbolic behavior is also possible with animals of lower relative brain size.

The Value of the Individual

Caregiving behavior is widespread in dolphin species.[95] What is remarkable is that it extends to those who are not kin and even to other species. Supporting behavior involves lifting a sick or wounded dolphin to the surface so it can breathe (as mammals, dolphins must take periodic breaths of air). The helping dolphin lifts from beneath, interrupting only to catch a breath itself, and remaining until the stricken animal recovers. This behavior has been observed in bottlenose dolphins as well as two other cetacean species.

This behavior is even extended to other species.[96] An Atlantic bottlenose dolphin was observed raising a stricken striped dolphin, and a pilot whale has been sighted carrying a dead striped dolphin. Female bottlenose dolphins have repeatedly been observed carrying dead calves on their rostra (front beaks), both in the wild and in oceanariums.[97] Zoologist Randall Wells, who has studied wild dolphins for more than 30 years near Sarasota, Florida, describes a female that lost her first calf a day after birth.[98] For nearly an hour and a half she circled, repeatedly lifting her dead son to the surface, whistling, dropping him again, then repeating this process every few minutes. This continued until the aggressive actions of two young males chased her off. This "mourning behavior" recalls the attachment of chimpanzees and elephants to their stricken young.

A variety of cetacean species are known to "stand by," imposing themselves between a hurt animal and a perceived threat.[99] This behavior is well known to whalers who try to improve their take by luring schoolmates to a stricken member. Bottlenose dolphins maintain an interest in their own kind, both kin and nonkin. The birth of a new calf often results in visits by older siblings that separated from the mother much earlier. Wells notes that babysitting of young dolphins may be undertaken by kin or other members of the band, both male and female.[100] One report notes that when a mother was giving birth, other dolphins formed a tight pack around her to protect her from sharks that were attracted by the blood.[101] Other dolphins herded the sharks away.

The curator at Marine Studios in Florida reports that during experiments with bottlenose dolphins, a young male received an overdose of Nembutal and developed difficulties in swimming.[102] Two females supported him at the surface for more than 20 minutes until he was able to swim. Only one of the females was believed to be related to the male. A report from the wild describes an incident in which a charge of dynamite was exploded near a school of dolphins, stunning one of them. Two adults swam to his assistance, supporting him from beneath so that he could get air at the surface. When the two helpers needed to catch a breath, other adults came to take their places. This caregiving persisted until the injured animal could swim on its own, and then the whole school left the area.[103] While these examples of helping behavior are based on anecdotes, which are generally viewed skeptically by scientists, Conner and Norris, who collected these reports extensively, say the evidence "is so common as to be overwhelming in its broad detail."[104]

Self-Recognition in a Mirror

Dolphins are among the few species capable of recognizing themselves in a mirror. In one test, two dolphins were marked with nontoxic black ink on a part of the body they could not see without the use of a mirror.[105] At other times, they were "sham-marked" in the same way, but with a marker that left no visible mark. The dolphins spent significantly longer time in front of mirrors looking at themselves in areas where they had been touched when they had real marks than when they had sham marks. Bottlenose dolphins have the special forebrain neurons known as *VENs*, which have been implicated in the existence of self-consciousness and social awareness.[106] VENs are found in similar forebrain areas in elephants, chimpanzees, humans, and dolphins, and these areas may be the seat of a higher form of consciousness about the self. These ideas will be addressed further in chapter 7, where I discuss hierarchical circuits in the brain.

All four of the species that possess the ability to self-inspect in front of mirrors consist of high-EQ animals that may have, as neuroscientist Lori Marino suggests, "a physical sense of self, or 'I,' robust enough to be conceived of and mentally contemplated."[107] In humans, mirror self-recognition does not emerge until about 18 to 20 months of age and corresponds with a time of increasing self-awareness and awareness of others.[108]

We have examined four peaks of brain development in different lineages and seen that they have a variety of qualities in common. These species have long-term relationships and complex societies with "politics" in which two or more individuals may form alliances against others. They appear to consider what others are thinking and can form flexible plans of deception. Hunting can involve different roles for different members that are coordinated with each other. They have complex communication with the possibility of understanding syntax in learned instructions.

Each of these species has a long period of juvenile dependency that allows for a slowly maturing nervous system capable of flexibility and innovation into the adult years. The variability of learning creates individual differences among members of a species and local cultural differences among groups. There is insight learning that finds quick solutions to novel problems faster than we would expect from trial and error alone.

At one level, we can simply say a more complex nervous system leads to more complex behavior. But I would like to go further and suggest that there is an emergent self that seeks to express itself in more complex communication and social relationships and that values others of its kind more fully as individuals than do species with smaller relative brain size. We see this in long-term relationships, what looks like mourning at the death of a conspecific, in helping behavior toward nonrelatives, and in the formation of higher-order coalitions in society. In all four groups, evidence of self-recognition in a mirror fits this picture of an emergent self that views itself and others with greater objectivity.

This emergent self also expresses itself in intricate manipulation of its environment. Each species has high dexterity, but the appendage employed—a hand, a beak, a trunk, or a flipper—varies widely according to the evolutionary origins of the species. What is constant is a large brain that is expressing itself.

This large brain also appears capable of abstraction. Each of these species is skilled at mimicry, and it could be argued that this involves a concept of the other that serves as a model. It suggests that there is a mental representation of some part of the model, such as its voice or movements, that the individual is trying to match. There also appear to be mental representations in tool making. The termite fishing sticks of chimpanzees and the leaf tools of the New Caledonian crows are made to standardized sizes that suggest preset concepts of length and width. They may then be carried long distances where they are just the right dimensions for the task at hand. This suggests a mental representation of both tool form and the goal of its use.

An emergent self sees with greater objectivity, and it can mix and match categories with greater flexibility. It can combine tools in novel ways and formulate new alliances in social life. It seems to view itself with greater objectivity, as evidenced by self-recognition in a mirror. In the next chapter, I will suggest a physical basis for this greater objectivity in the hierarchical arrangement of circuits in a large brain.

The animal portraits in this and the previous chapter could be extended to other species that have large relative brain size, such as parrots (Psittaciformes), which are an entirely different lineage from the songbirds; orangutans, among the primates; and the largest-brained invertebrates, octopi and squids (Cephalopoda). We would see much of the same suite of characteristics I described for the four animal groups here, although any one species may not have all these characteristics. I have chosen these four species because they have the most complete behavioral studies in the wild, and they represent such different lineages and habitats.

Overall this might be described as an evolution of personality, for these species are more fully "persons" in the human sense. They have more to express to each other and in their handling of their world. They are more distinctly individual and value each other as such. I do not wish to appear rampantly anthropomorphic, but instead to suggest that there may be a constellation of qualities that are involved with each other. Long-lasting relationships, complex communication, flexibility, and inventiveness are qualities of a large, slowly developing nervous system. Humans are one peak of this evolutionary development, but we also see it emerging in other lineages, in other families of mammals as well as birds.

All of these species are K-selected, equilibrium species. They share the qualities of large brain size, long dependency, and long life span. They are capable of extensive learning and flexible behavior that allows them to adjust to changes in their environment. We saw this earlier as one of the fundamental strategies of life in dealing with a world of constant change. Life either produces many organisms of low information content that mature quickly to find the variants best suited to new conditions, or it produces fewer organisms of high information content and wide behavioral flexibility as a way of coping with change. R-selected opportunistic species and K-selected equilibrium species are end points of a continuum that describes fundamental strategies of life in dealing with a world in constant flux. Humans are currently, on this planet, at one extreme of K-selection, having extended information gathering to languages, books, computers, and the Internet. But the rudiments of culture and language exist in other species around us, from the cultural differences in tool making of chimpanzees and corvids to the comprehension of syntax in dolphins and chimpanzees. We see a common strategy laboring to express itself through different lineages. Had it not come to dominance in the primate order, it might well have done so in one of these other lineages. It might even be inevitable as part of the evolutionary process wherever it has had time enough to run.

Life has a drive toward homeostasis, to buffer itself against changes in the environment and maintain conditions optimal for life. Behavioral versatility, having many possible reactions to change, can be seen as one form of homeostasis. Alternative programs can be built up in either genes or nervous systems. On any world where life has begun, this drive toward homeostatic stability is likely to happen. Chapter 11 addresses some of the latest findings in astronomy that indicate that habitable worlds may be found all over the universe. The pageant of life is so diverse on our own planet that we cannot predict the shape life would take on other worlds. But I think it can be forecast that behavioral versatility will be one of the fundamental coping strategies of life anywhere. This will require complex programs that need long development, and therefore probably extended parental care. Such an organism will probably have long-term attachments and know something of the emotion we call love. It will also probably have complex communication, even if it be in modalities other than sound, and it will likely express itself in complex manipulation of its environment, although it may be through very different appendages, such as an evolved limb, flipper, beak, or trunk. The supreme product of the evolutionary process may be a person, an emergent self with complex relationships and communication and a higher level of objectivity on self and its surroundings.

7

Concepts as Feature Extraction

An emergent self that can view others, itself, and its environment with greater objectivity may be the result of a large brain that has extra neurons that can be dedicated to higher levels of abstraction. In this chapter, I will describe an arrangement of brain circuits that allows one set of neurons to abstract qualities from responses in another set of neurons. Recall that an animal with an encephalization quotient (EQ) above 1 has higher brain weight than an average animal of its body weight. This extra brain matter may be available for functions beyond mere physiological control. Also, as brains become larger in absolute size, an increasing percentage of the neurons appear to be dedicated to higher functions.[1] Thus an animal with a small body but high EQ (such as a corvid), or a large animal with a large brain but not especially high EQ (such as an elephant), may both have many extra neurons available for additional analysis of the raw data coming to the brain.

Neurons as Calculators

Concepts are abstractions, and as the term implies, they depend on lifting details from a larger field. To do this, you must in a certain sense be "above" the field so that you can note aspects that share similar qualities, much as you have to be above the forest to see a pattern in the trees. From the study of various round objects—for example, a peach, an apple, and a cherry—you might abstract the concept of a circle. A neuron is ideally suited to respond to aspects of experience for it can take many inputs (in the thousands) and render a yes or no decision on a summation of their influences. To understand this, we need to explore the basics of a neuron as a summation device.

A neuron is a cell like any other cell of the body, but one that has specialized for handling electricity. Our nervous system and brain consist of billions of these cells, but they are much more than the wires that connect parts in our electronic devices. They can both conduct electricity and sum electrical inputs, so they are really microcalculators, more like a computer chip than a copper wire.

Figure 7.1 shows an outline of a neuron that I will call *neuron 1*. Like all cells, it has a nucleus in its center and is covered with a plasma membrane. The cell has a long process that extends outward, known as its *axon*. I have indicated a "spike" of electricity that would pass down the axon if this neuron depolarized. The membrane extensions near the

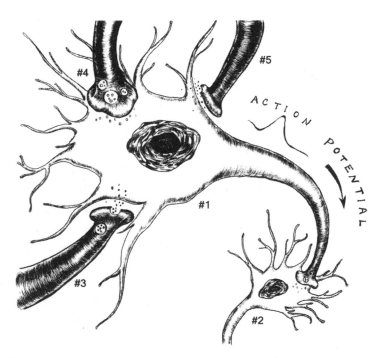

Figure 7.1 Neuron 1 is influenced to fire by input from neighboring neurons. Neuron 1 receives input (neurotransmitters) from neighboring neurons (3 to 5). Their sum total causes neuron 1 to fire an action potential, leading to the release of neurotransmitter onto neuron 2.

nucleus are called *dendrites*, and I have indicated where axons from three other neurons (3 to 5) are terminating near the dendrites of neuron 1. An actual brain neuron may have thousands of these inputs from neighboring neurons, but three are enough to illustrate the basic principles.

Near the end of each axon are vesicles (tiny membrane-covered vessels) of neurotransmitter, a molecule manufactured by this cell and stored near the end of its axon. When a spike of electricity passes down the end of a neuron, it will cause these vesicles to fuse with the end of the axon and spill their contents onto the next neuron (2) in a series. Much of modern psychiatry is focused on this little operation, for neurotransmitters like dopamine and serotonin have important effects on groups of neurons that influence our moods.

The membrane of every neuron is studded with tiny pumps (not shown) that are specialized for pumping ions. Most of these are differential pumps that transport three sodium (Na+) ions out of the neuron for each two potassium (K+) ions they move inside. Thus, overall there is more "plus" charge outside a neuron than inside. In electrical terms, you can say the interior of a neuron is negative with respect to its exterior environment. A neuron at rest is at about −70 millivolts (mV) in relation to its outside. This is a tiny differential, but enough to give us shocking experiences if we bump our elbows or knees in the wrong way. An ordinary flashlight battery at 1.5 volts has more than 20 times the voltage differential of a neuron.

Every neuron makes a yes or no decision, to depolarize or not. If its interior becomes more positive, rising from −70 mV to about −55 mV, it "spikes," sending a wave of depolarization down its axon (see figure 7.1). Anything less than this change (say from −70 to −60), will not provoke a depolarization, and the neuron can also become more negative (say, from −70 to −80), in which case it is even less likely to spike because it is hyperpolarized.

What can change the interior of a neuron? Here is where the influence of its neighbors becomes important. When neuron 3 spills its neurotransmitter, that molecule can affect ion channels in the membrane of neuron 1 (not shown). It might open Na+ channels, allowing the higher concentration of Na+ outside neuron 1 to flood back into the cell, raising its interior to −55 mV (or even more positive), in which case neuron 1 spikes. Suppose neuron 4 spills a different neurotransmitter that affects another kind of ion channel, say a K+ channel. In that case, the higher K+ concentration inside neuron 1 causes K+ to rush outward, making neuron 1 more negative, so it does not spike. In fact, the effect of both neighbors could cancel each other out, one making their target cell more positive, the other more negative, and the sum would not budge neuron 1 past its resting stage.

In actual fact, a typical neuron in the human brain has about a thousand inputs of this kind on its highly branched dendrites, and it can have up to 100,000 connections (synapses) with neighbors.[2] The complexity is almost beyond imagination, yet a simple calculation is always made. If the sum of all the ions entering and leaving the target cell raises it to −55 mV, it will fire; if that magic number is not reached, the neuron remains at rest. Cells are small because they rely on diffusion to accomplish many of their tasks. Think of the tiny cell body of the neuron like a stew pot, surrounded by many chefs, each throwing in his particular spice. All this input mixes very quickly in the tiny volume surrounding the nucleus, and if the electrical total reaches threshold, the neuron fires. If it all adds up to something more negative than −55, it will not. In this way, every neuron can be a microcalculator, summing many influences and making a simple yes or no decision on the total.

This ability to calculate allows neurons to form circuits that focus on particular patterns of activity in their neighbors. A master neuron can make a decision on inputs from subsidiary neurons. This principle can lead to hierarchies of circuits that derive ever-more abstract qualities from the raw data. I will argue that an animal with high relative brain size, with its extra neurons, will be able to build additional levels in this hierarchy, leading to ever-more abstract understandings of experience.

Visual Circuits

The place where hierarchical connections are best understood is in visual circuits, which probably serve as a model for mechanisms found all over the brain. It is said that the eyes are windows to the soul, and in fact the retinas in the back of our eyes are made of neurons that have migrated directly out from the embryonic brain to near the surface of the skull in order to participate in forming our eyes. Figure 7.2 shows the circuits of four layers made by neurons of the human retina. At the base are the photoreceptors, the only cells actually capable of capturing light. There are two types: rods and cones. Cones are for color, while rods see only in black and white, although they are about 1,000 times

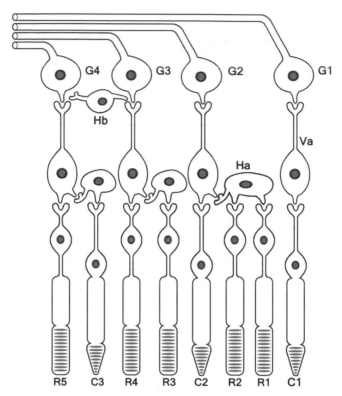

Figure 7.2 Simplified view of the connections between neurons in a portion of the retina of the eye.

Rods (R) and cones (C) respond directly to light. Information is sent back to the brain by way of ganglion cells (G1 to G4). The ganglion cells are connected to different kinds of circuits. At the far right, ganglion cell G1 is reporting the output from a single cone cell (C1) by way of another vertical neuron (Va). Elsewhere, two horizontal layers of neurons (Ha and Hb) form more complex circuits. (From Neil A. Campbell and Jane B. Reece, *Biology*, 6th ed. [San Francisco: Benjamin Cummings, 2002], 1068. Adapted by permission of the publisher)

more sensitive to light than cones. The highest density of cones is in a central region of the eye known as the *fovea*. We see with the greatest detail in this region because many of these cones report directly to the brain. Much of our eye movements are a matter of turning our foveas toward areas of interest. As you read this page, for example, your foveas are focused on individual words. Look at one word and at the same time try to discern words at the periphery of the page. You will see how much more acuity you have in an area of interest. This is not just a matter of attention, but the one-to-one connections made between a cone in the fovea and the neurons that report back to the brain.

The circuit for one cone in the fovea is shown at the far right of figure 7.2. Note that cone C_1 is directly connected to another vertically oriented cell (Va) that connects to a ganglion cell (G1), whose output goes directly to the brain. Thus in the fovea there can be a nearly one-to-one relationship between the cones receiving light in the retina and the ganglion cells reporting back to the brain.[3]

The eye does not always emphasize such great visual acuity. Elsewhere in the retina, horizontal cells may pool the output of many photoreceptors before they converge on a ganglion cell. Two such circuits are shown at the left in figure 7.2. Outside the fovea, horizontal cells (Ha) are pooling the output of rods and cones and influencing the firing of the next vertical layer of the circuit (Va, technically known as *bipolar cells*). Note that ganglion cell G_2 is receiving pooled information from rods R_1 and R_2 and cone C_2. It is therefore getting more input from light but not responding with the single-cone detail of ganglion G_1. This is perhaps the simplest example of how different neuron architectures can emphasize different aspects of vision.

Information gathered at the second vertical layers of cells (Va) may be further combined by cells at a second horizontal layer (Hb, technically known as *amacrine cells*). Note that Hb is linking ganglion cells G_3 and G_4 and thus combining output from rods R_3 to R_5 and cone C_3. Figure 7.2 simplifies connections found in the retina, but the basic principles of combining neurons to emphasize particular aspects of vision are essentially the same everywhere in the eye.

At night, for example, horizontal cells (the layer labeled Ha) pool the output of many rods to provide night vision. Despite low levels of output from the photoreceptors, their signals can be combined to stimulate a ganglion cell. Thus, at night we do not see with the detail of daytime vision, but we are able to see in light that may be billions of times dimmer than daylight.[4]

Different circuits in the eye are specialized for different aspects of vision. In the fovea, you see with great acuity but are less sensitive to brightness or overall luminance. Outside the fovea, you can see in dimmer light, but with less detail. You can test this on a nighttime walk. Look at a bright star or the moon. At the corner on your eye, pick out a dim star. Now focus on this dim star directly, and it will look much dimmer, or disappear. Your central fovea is not as sensitive to luminance as the peripheral areas of your retina.

Each retina has about 125 million photoreceptors but reports to the brain with only 1 million ganglion cells. Thus, overall, only $1/125$ of the information from the primary response to light by the rods and cones is sent back to the brain. Different kinds of circuits pool information in different ways to focus on different features of vision, with a ganglion cell sitting at the top of a decision tree. Eyes did not evolve to be accurate television cameras. They evolved to focus on those aspects of vision most essential to survival.

Different patterns of circuits pick out different aspects of visual experience. Some of the ganglion cells are selective for direction. There are different ganglion cells for each of the cardinal directions: left, right, up, or down. The second layer of horizontal cells (Hb) appears to play a crucial role here. A ganglion cell that fires to motion in one direction is inhibited from firing by an object moving in the opposite direction. If ganglion cell G3 in figure 7.2 is sensitive to motion from right to left, it would receive an inhibitory neurotransmitter from the amacrine cell in layer Hb for an object that moved from left to right. In general, for a given ganglion cell, motion in the null direction provides an inhibitory signal first, then a stimulatory signal, whereas motion in the preferred direction has a stimulatory signal followed by an inhibitory one.[5] Thus the timing of signals as well as circuit architecture can influence the result at the top of a decision tree. And neurons, as we have seen, are masters of integrating information, able to total up multiple influences and make a yes or no decision. The number of inputs, as well as the ratio of excitatory and inhibitory influences, will determine the result. Many such single-unit decisions can add up to responses that are quite sophisticated.

Visual Processing in the Brain

The four layers of circuits found in the retina appear to be a model for what happens elsewhere in the cortex of the brain. Here also, neurons are arranged in layers. The cortex is traditionally described as being six layers thick, although at least eight different cell types have been identified in the upper section of the brain, known as the *cerebral cortex*, a rind of tissue only 2 millimeters thick. In an irony of wiring, vision received by the eyes at the front of the head is processed by the brain at the back of the skull, and the left side of the visual field is evaluated on the right side of the brain, and vice versa. Columns of neurons (running perpendicular to the surface of the skull) appear to share common characteristics, such as the direction of an object's motion or its angle of orientation.[6]

The process of abstracting qualities from experience continues in the visual cortex of the brain. It provides insights into how ever-more abstract qualities can be derived by neuronal architecture. The first area that analyzes information from the eyes in the back of the brain is known as *V1*, and rising above it, like terraced gardens on a hill, are areas of further processing called *V2*, *V3*, and *V4*. As you rise in this hierarchy, groups of cells are dedicated to ever-larger receptive fields of view and can respond to greater image complexity.[7] Neurons in V1 are sensitive to bars of light in a particular orientation, but neurons in V4 respond to more complex shapes: lines with convex or concave curves. These, in turn, can be used to build up more complex images, much as the letters of an alphabet are used to make many words.[8] The output from V4 goes to the temporal lobe, an area of the brain near our ears. Neurons here respond to faces, and it has been theorized that facial shapes are put together from the lines and curves processed earlier in the visual hierarchy. Individual neurons have been found that respond to pictures of familiar faces, even celebrities. In one patient, for example, a single neuron fired at high rates in response to photos, even cartoons, of Bill Clinton seen either "face-on" or in profile. Pictures of other people, animals, or buildings provoked much lower firing rates.[9] It is absurd to think there is an individual neuron tuned to Bill Clinton. But there could be a neuron that sits at the top of a large decision tree, getting inputs from many other neurons that respond to lines and curves that define a particular shape of face, smile, and style of hair, and if the right number of these inputs are combined, they might cause a summary neuron to fire. Remember, a single brain neuron can receive thousands of inputs from other neurons, much as a lot of pixels can define a picture.

The main point is that the neuron as a summation device can abstract patterns from more complex data. It can form circuits that selectively respond to particular shapes or types of motion. A neuron can sit at the top of a hierarchy and decide when the right pattern has been activated at a lower-level circuit. These circuits in turn can be used as subunits in another hierarchy that extracts ever-more abstract qualities from the raw data. We have not yet traced all the connections that make up these higher circuits, but it is likely that they are elaborations of mechanisms seen earlier, all the way back to the retina of the eye. Nature often reuses effective mechanisms and combines them in new ways. There are important families of genes, for example, built up as repeats of a single ancestor sequence that has been modified slightly to take on new functions. The genes for different types of color vision are like that, as are the genes for capturing oxygen in the blood, and even the sequences that determine the identity of body parts (the Hox genes). Nature is not so much an inventor as a tinkerer, recycling old parts and trying them in new combinations to find new functions.

Neurobiologist Semir Zeki has pointed out that the brain is always conceptual.[10] It imposes categories on experience because the raw data are too diverse and chaotic to absorb as a whole. It focuses on those aspects of experience that are most essential to survival. Cells in V1 that are selective for orientation do not care if the object is blue or red, a stick or a pencil; they fire only in response to objects at a certain angle of orientation. Other neurons that are specific for motion respond only to the direction of motion and will do this equally for a car or a ball. Neurons that process color vision also follow a conceptual framework. They are concerned with the three primary colors: red, blue, and green. A leaf will look green at dawn or at midday, even though the actual amount of green light reflected from its surface may be very different in bright light than in dim. The visual cell for color is actually responding to a ratio of wavelengths from the surface of the leaf in relation to its surroundings, which, if they are nonliving matter, are probably reflecting more reds and blues. The vision cell is imposing a concept: if the ratio of green versus the other two colors is above a certain threshold, the leaf is labeled green. The greenness does not exist anywhere in nature; it is a label created by the brain.[11] This calculation takes place in V4, and if these neurons are destroyed, the patient becomes colorblind, even though the original color receptors way back in the retina, the cones, are still functioning normally.

Frontal Areas

Higher derivatives of information are made in frontal areas of the brain, and it may be said roughly that the more anterior the region of the brain is, the more abstract the processing becomes. The anatomy of these areas provides insights to their function. Although humans have a high brain:body ratio for a mammal, the dimensions of our frontal lobes are not unusual for a brain of that absolute size. But the same is not true for areas within the frontal lobes. The foremost area, known as the *prefrontal cortex* (*PFC*), is unusual in multiple ways. It is disproportionately large: in cats it constitutes 3.5 percent of the total cerebral cortex, in macaque monkeys it is 11.5 percent, in chimpanzees 17 percent, and in humans it is 29 percent.[12] In primates, it has added an extra layer of neurons to the usual six-layered arrangement of the neocortex.[13] Subregions of the PFC are highly interconnected—with each other and with other lobes of the brain. They receive input from vision as well as the other senses, including touch. This input comes not from the senses directly, but from their higher processing centers, such as the final stages of visual processing already discussed. The neurons of the PFC are highly plastic, subject to training and the learning of rules quickly. Their activity is affected by the state of attention and motivation of an individual, and they can fire in anticipation of an expected reward. They output to premotor regions that prepare the body for action, and they are believed to have an executive function in mapping out goals and ways to achieve them.[14] The frontal lobes are also among the last to mature in human development, following a roughly back-to-front movement, with the PFC maturing last. It is interesting that the areas last to mature are also the most recent to increase in size in evolutionary time.[15]

Only primates appear to add a separate area to the PFC,[16] labeled *D* in figure 7.3. (This area is known technically as the *dorsolateral prefrontal cortex* [*DLPFC*], so by labeling it *D* I am simplifying an unwieldy nomenclature.) In humans, gray matter in the prefrontal cortex increases in volume after birth, reaches a maximum sometime between four and 12 years of age, and decreases gradually thereafter. This is part of the pruning involved in the "wiring by firing" process discussed in chapter 3. In contrast, the underlying whiter matter—composed of the highly insulated neurons that connect neurons of gray matter to each other—increases from childhood into young adulthood.[17] Compared with the apes, who are our nearest relatives in the family Hominidae, humans have 23.6 percent larger prefrontal lobes, but most of the difference is in white matter. We have 7.3 percent more gray matter, but 42.5 percent more white matter.[18] Thus there is an unusual amount of connectivity between different areas of the brain in our species.

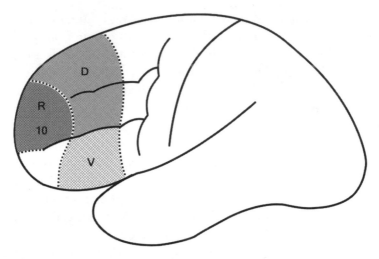

Figure 7.3 Three areas of the human prefrontal cortex.
The most anterior area R is involved in highly abstract thought.
(From K. Christoff and K. Keramatian, "Abstraction of Mental Representations: Theoretical Considerations and Neuroscientific Evidence," in *Neuroscience of Rule-Guided Behavior*, ed. Silvia A. Bunge and Jonathan D. Wallis [Oxford: Oxford University Press, 2008], 108. Adapted by permission of the publisher)

Of special interest is the most anterior area of the brain, labeled *R* in figure 7.3 (also known as *area 10* in Brodmann's older terminology). It is unusually larger in humans than in other primates, although there is controversy about how much larger it is.[19] This area has lower cell density than other areas of the forebrain but more connectivity between its neurons.[20] It thus has high "introspective" potential to pass information back and forth for complex analysis. It is implicated in some of the most abstract levels of thinking.

Higher Derivatives

Area R gets input from lower visual areas after they have processed whole images in the manner discussed earlier. It is able to compare and contrast categories of images to make relational decisions. For example, a brain imaging study looked at where the brain processed concrete rules as opposed to first-order and second-order abstract rules. Human subjects had to make increasingly complex decisions about the orientation of sets of partially filled circles. For the concrete and first-order abstract rule, area V of the PFC was active (see figure 7.3). For the second-order abstraction, areas V and R were active. The researchers suggest that area R becomes involved when higher-order rules have to be applied.[21] A similar result was obtained in tasks that involved unscrambling anagrams of words that were increasingly abstract.

Area R is also implicated in recognizing true analogies, such as "planet is to sun as electron is to nucleus." In the study, a true analogy activated area R, whereas a nonanalogy ("cow is to milk as duck is to water") did not.[22] As analogies became more abstract, this brain area became more active. Also, cross-domain analogies ("nose is to scent as antenna is to signal") recruited higher activity in area R than within-domain analogies ("nose is to scent as tongue is to taste").[23]

We are a long way from understanding creativity, but analogical thinking—being able to see the connections between disparate areas—is surely an important part of it. Think of August Kukulé, who discovered the ring structure of benzene after having a dream about a snake biting its own tail, or Charles Darwin, who reasoned from the artificial selection that he saw among animal breeders to the concept of natural selection in the wild.[24]

Although a number of researchers support models of hierarchical processing in the brain,[25] the issue is not yet settled, and it may be that disparate interconnected areas are involved in arriving at higher abstractions.

The third-order relations that we saw for high-EQ animals in the last two chapters may be the result of having extra brain regions that can take higher derivatives of information. As mentioned before, this can be thought of as a forest-and-trees problem. While you are within the forest, it is hard to see patterns of how the trees are arranged, but viewing the forest from above, you might more easily discern particular groupings. Hierarchically arranged circuits of neurons, where one circuit is tuned to particular patterns of activity in a lower circuit, could provide just such objectivity. Animals with high relative brain size may have "extra neurons" beyond those needed for basic physiological maintenance that could be specialized for such data mining, and they may also have more brain modules that could be devoted to additional levels of processing.

Special Forebrain Neurons

The previously discussed Von Economo neurons (VENs)—found uniquely in forebrain areas of humans, dolphins, elephants, chimpanzees, and most of the other great apes—are large spindle-shaped neurons believed to convey information quickly between different forebrain areas. In all these species, they are found in similar frontal areas and are more numerous in the right hemisphere of the brain than the left.[26] In humans, they are found in both areas D and V of the prefrontal cortex (see figure 7.3).[27] In humans and other primates they are also found in another frontal area known as the *anterior cingulate cortex*.[28] In both humans and chimpanzees, these neurons first appear late in gestation, and in our species they are more numerous than in any of our primate cousins.[29]

The function of these neurons is not entirely clear, but their involvement in some human neurodegenerative disorders is intriguing. In frontotemporal dementia, there is a loss of self-awareness and the ability to understand the intentions of others. People lose emotional awareness, empathy, and the ability to view the self in relation to others. It involves a 74 percent reduction in VENs and affects the right hemisphere of the brain more than the left.[30] VENs also degenerate in Alzheimer's, and a forebrain area with these neurons is reduced in autism.[31]

The species in which VENs are present have complex social lives and recognize themselves in mirrors. In humans, degeneration of prefrontal areas that have VENs results in an impairment of the sense of self and its relationship to others socially. These frontal areas may be the location for processing a higher, more abstract sense of self. But what exactly does this mean? All species must have at least some sense of self because they are not constantly bumping into things as they move about. There must be some rudimentary sense of where the body ends and the external world begins. But this ability can be given to simple robots. It takes only a sensor, a motor to power motion, and a program to change the direction of motion when the sensor reports that other objects are coming too close. No one suggests such a servomechanism has self-consciousness.

But what of a brain that could view others and the self like pieces on a chessboard and make decisions about relations between them? Such consciousness would require a kind of objectivity. While you are within a maze, it is hard to decide which way to turn, but when viewing the maze from above, it is easy to see the pathway leading out. A brain that could make representations of both the self and others and had extra circuits to decide on relations between them, like a player looking down on pieces of a chessboard, would have the required objectivity and versatility. A hierarchical arrangement of circuits, with frontal areas being able to compare and contrast information from earlier circuits, might have such abilities, and a high-EQ brain might supply the extra neurons to build such circuits. Also, any one of these factors might be lost in a neurodegenerative disease, leading to a diminished sense of self, others, or the relationships between them.

Brain Modules

As brains become larger, they become more modular. There has been a general increase in the number of distinct areas in the forebrain during vertebrate evolution that parallels increases in relative brain size. Frogs are estimated to have about 50 distinct cell groups in the brain; reptiles, 85; birds, 87; and mammals, 257.[32] Neurobiologist Georg Striedter suggests that this evolutionary increase often follows a pattern where, as a particular area increases in size, neurons at its far end may begin to take on a new function. These ultimately grow and differentiate into an island of neurons specialized for a distinct purpose. Thus a new vision area like V3 evolves next to an old vision center, V2,[33] and we get the steplike hierarchy of vision processing outlined earlier.

Brain modularity follows from the geometry of how vertebrate brains typically grow. We can see this in the growth of the neocortex, the six layers of neurons that compose the upper rind of mammalian brains. As the brain matures, the neocortex increases in volume at a faster rate than other parts of the brain.[34] This is true not only in the growth of individuals as embryos, but in evolution when we compare mammal species with different relative brain sizes.

The surface area of the neocortex varies over 1,000-fold between different species of mammals, but its thickness does not vary more than about sixfold.[35] This difference in thickness is mainly due to white matter, neurons that are heavily insulated with fatty membranes and that serve to interconnect the neurons of gray matter above them. This means that as a brain becomes larger, surface area is increasing at a much faster rate than the connections between different areas. The result is that the brain becomes modular, condensing into local areas that process particular aspects of cognition and then sending their results onward to the

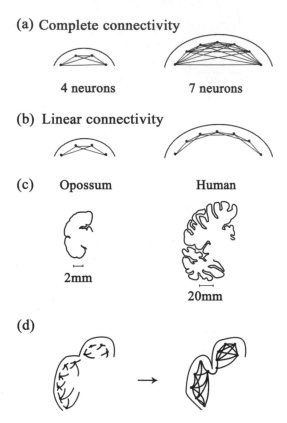

(a) Complete connectivity

4 neurons   7 neurons

(b) Linear connectivity

(c)   Opossum   Human

2mm

20mm

(d)

Figure 7.4 Geometry of brain growth.
(a) As brains become larger (e.g., from four to seven neurons), if each neuron maintained complete connection to every other neuron, the number of connections would increase exponentially. Instead (b), each neuron maintains a limited number of connections to neighbors. (c) The brains of high-EQ mammals are more highly folded, thereby increasing their surface area. (d) Outward folds result in areas that have more interconnections. (Portions [a] and [b] based on data in Georg F. Striedter, *Principles of Brain Evolution* [Sunderland, Mass.: Sinauer, 2005]. Portion [d] based on data in D. C. VanEssen, "A Tension-Based Theory of Morphogenesis and Compact Wiring in the Central Nervous System," *Nature* 385, no. 6614 [1997]: 313–318)

next module. One can see this in the geometry of brain growth in figure 7.4a and b. As a brain becomes larger with more neurons, if they maintained complete connectivity to each other (figure 7.4a), white matter would increase exponentially faster than the gray matter of the cortex above it. But white matter increases only slightly faster than gray matter in the average mammalian brain because a typical neuron (outside the human PFC) maintains about the same number of connections to other neurons, regardless of brain size.[36] I have called this *linear connectivity* (figure 7.4*b*). The arrangement creates local modules of processing that send their results on to the next module. Such a system becomes hierarchical, a kind of assembly line of information, where parts are gradually put together into larger wholes, as we saw with the processing of vision. Also, our prefrontal cortex, as mentioned, is unusually interconnected with high levels of white matter, which goes through late maturation in early adulthood. In chapter 8, I will discuss an important influence this delayed maturation may have had during a seminal period of human culture.

Connections between brain areas also help explain the folding pattern of brains, which increases in mammals with larger brain size (figure 7.4*c*). Connections between areas of the cortex are established early, and an area with many interconnections will cause the cortex to bulge outward (a gyrus) as the young brain proceeds to grow (figure 7.4*d*). Alternatively, weakly connected areas with sparse numbers of neurons between them tend to be separated by an inward fold (a sulcus), as the brain matures.[37]

This exercise in brain geometry is supported by detailed studies of connection patterns between neurons in brains. Modern staining techniques have allowed researchers to trace the connections between some 16,000 single neurons in the fruit fly's brain. The adult fly's brain contains about 100,000 neurons in total, but this sample may be enough to outline its basic wiring diagram. Researchers found 41 local processing units, functional modules whose neurons have many interconnections to each other and are separate from other such modules. A distinct group of projection neurons connects one module to another.[38]

This pattern follows a model that has long been hypothesized for human information processing. The human cortex contains minicolumns, groups of 80 to 100 neurons that span all six layers of the cortex and have many interconnections to each other. They may form a processing unit, much as the five layers of neurons in the retina build individual circuits focused on some aspect of vision. Minicolumns, in turn, are bunched together into larger cortical columns, which may be devoted to a particular task of information processing. There is much internal connection within columns that is distinct from other such columns, and many distinct neurons send the output of one column on to another column.[39] These may be acting like so many desktop computers linked together to increase their overall computational power.

Modular organization of the brain allows both linear and parallel processing. That is, one set of modules may be processing one set of data while another set is processing a different stream of data at the same time. The brain gains computational power from both hierarchical processing in a single stream and comparison of information between parallel streams.

Finally, it should be noted that not all species have the highly folded brain surface of mammals. Birds have a smooth cortex, yet they can have very advanced cognitive abilities, as we saw with the corvid family. Both groups evolved large brains by increasing the size of the embryonic forebrain, although birds enlarged a different area than mammals did. The area in birds, known as DVR, has neurons in layers, although they are oriented in a different way than the layers in the mammalian neocortex.[40] Although their bodies are generally smaller, birds like corvids and parrots can have EQs similar to those of highly encephalized mammals (see figure 3.4).

The operation of a neuron, and the architecture of the way neurons are connected in brains, can explain many of the qualities we saw in large-brained animals in the last two chapters. Neurons are ideal summation devices, able to take many inputs and make a decision on the totality of those influences. They can sit at the head of a decision tree and determine whether a set of criteria has been fulfilled. In this way, they can abstract qualities from the raw data, such as whether there is sufficient luminance, or what direction an object is moving in. Circuits can be built on top of circuits in hierarchical fashion, deriving information in an ever-more abstract fashion, the way a face can be put together from a set of lines and curves.

Animals with high EQ, or with large absolute brain size, have more extra neurons to devote to higher cognition, beyond mere mechanical control of the body. Total connectivity does not keep up with the proliferation of gray matter in such brains, so they divide into modules, specific groups of neurons devoted to a particular processing task. Large brains have more modules, and these may be the source of the third-order relations seen in species with high relative brain size.

The full pathways to higher abstractions have not yet been traced, but it is reasonable to assume that they are extensions of the mechanisms discussed here. A hierarchical arrangement of circuits could explain the third-order relations seen in large-brained animals using a tool to modify other tools, or forming complex associations in the "politics" of group life. It takes a certain mechanical objectivity to stand back and look at a rock and consider how it might be modified by another rock to make a better anvil or tool. It takes a certain social objectivity to consider how an alliance with one individual might be used against another individual. That objectivity might be supplied by a circuit that stands above another circuit and combines its salient qualities. Syntax in language can also be considered a kind of third-order relation, because it defines the relationship between words. Higher objectivity is also implied in mirror self-recognition, an ability rare in the animal world. Finally, behavioral flexibility may result from the ability to contemplate a variety of courses of action and choose the best one.

In this respect, human evolution, with its increasing brain size, can be seen as the continuation of a long-term trend in a variety of animal lineages. Our increasingly sophisticated tools, social organization, arts, and language may be the result of hierarchical arrangements of circuits that can take ever-more abstract derivatives from lower levels of processing. An eagle, after all, can solve simultaneous differential equations, although it does not have the symbols to represent them. When the bird swoops down to catch a fish in a stream, it must adjust its rate of change to the rate of the swimming fish, and it can predict the point of intersection, as evidenced by the meal it carries off in its talons. Solving simultaneous differential equations for their point of intersection is a notable achievement for a first-year calculus student, but the eagle does it intuitively. It just does not have the circuits to describe what it is doing.

Neuron Counts and the Cerebellum

Neurobiologist Suzana Herculano-Houzel and colleagues have pioneered a new way to count neurons in brain tissue. Using an antibody to label neuron nuclei, they have obtained what are probably more accurate counts of neuron numbers in specific volumes of brain tissue than earlier methods provided.[41] The results are in many ways surprising: (1) The cerebral cortex, which makes up 82 percent of the total brain mass in humans has only 19 percent of all the neurons in the brain. The cerebellum, which grows out from the hindbrain, has only 10 percent of the brain's mass, but 80 percent of all brain neurons.[42] (2) The cerebral cortex and cerebellum vary together in numbers of neurons across four orders of mammals, comprising 19 species, including humans. The two brain areas appear to be a functional ensemble.[43] In comparisons of different species of mammals, as brains become larger, an increasing number of neurons are found in the cerebral cortex plus cerebellum compared to the rest of the brain. Both brain mass and the numbers of neurons increase in a linear fashion with increasing body mass, but at different rates in different mammalian orders. Primates add neurons at a faster rate as brain size increases than do rodents or insectivores.[44] Thus a gram of primate brain has more neurons than a gram of rodent brain.[45]

The cerebellum grows out from the embryonic hindbrain while the cerebral cortex develops from the forebrain (see figure 3.5). Yet the two areas maintain reciprocal connections with each other and appear to have evolved in concert, enlarging together in similar proportions. Prefrontal cortex has increased 4.43 times in size and cerebellar cortex has increased 3.08 times in comparing humans with chimpanzees. Areas within the cerebellum that connect to the prefrontal and motor cortex have increased from about 57 percent of the cerebellum in capuchin monkeys to 67 percent in chimpanzees to 84 percent in humans.[46] The cerebellum integrates sensory and motor functions and is involved in visual problem solving and the learning of procedures. These higher functions appear to be localized in two lobes (the lateral hemispheres), which have many folds and a modular organization, similar to what is found in the two larger hemispheres of the cerebral cortex.[47]

Herculano-Houzel suggests that absolute brain size in terms of neuron numbers in the cerebral cortex and cerebellum is a better index of higher cognitive powers than values like EQ. She notes, for example, that even though humans have a high brain:body ratio in terms of mass compared to animals in general, the human brain is not unusual for a primate brain of it size. It has about as many neurons as we would expect of a generic primate brain of its weight (1.5 kilograms [3.3 pounds]). But it is the largest primate brain, and since primates have especially high neuron densities, we may have the most brain neurons of any species on Earth.[48] Elephants and whales, with larger brain volumes, appear to have lower neuron densities, so their total neuron count may be lower than ours, though direct measurements of neuron numbers have not yet been done for these species with the new technique described earlier.

There are problems, however, in taking total neuron counts as an index of cognitive abilities. As described in chapter 5, chimpanzees and corvids converge on many similar mental characteristics in terms of tool making, complex communication and social relations, innovation, imitation, and mirror self-recognition. A review describes a similar mental "tool kit" in corvids and apes.[49] Yet in terms of brain size, the chimpanzee brain at 390 grams (13.8 ounces) compared to the raven at 14 grams (half an ounce) suggests there should be some 28 times the mental abilities in the chimpanzee. The brain:body ratio for mass, however, is similar in chimpanzees and crows, with significant enlargements of the forebrain in both species. The New Caledonian crow (*Corvus moneduloides*) was cited in chapter 5 for its extraordinary tool-making ability and general learning skills. It has been observed in third-level tool use in employing one tool to obtain another. Yet its brain averages only about 7.56 grams (about ¼ ounce), though it has a high EQ for a songbird.[50] It has significant enlargement in forebrain areas devoted to motor learning and association compared to other birds.

The problem with simple neuron counts becomes starker when considering some of the advanced capabilities of some insects. The honeybee (*Apis mellifera*) is capable of rule learning and categorization, and has something very close to symbolic language in its waggle dance, by which it communicates the distance and direction of food rewards to nest mates. The honeybee brain has fewer than a million neurons, with two lobes known as *mushroom bodies* that mediate learning and memory. These bodies each contain only some 170,000 cells, but they are enlarged in bees and other insect species that show high levels of learning and flexible behavior.[51] As mushroom bodies become larger, they have more distinct modules of neuron organization, similar to what happens to the vertebrate cerebral cortex as it becomes larger in different species. A study found 12 modules in the honeybee brain compared to six in a moth.[52]

Finally, cell numbers may not be the best way to assess the processing power of particular areas of the brain. There can also be differences in neuron size and the numbers of synapses between neurons. Neuron densities can vary as much as five times in different cortical areas. A study of vision processing in the primate brain found much higher neuron densities in V1, the first area that receives information from the eyes than in higher vision centers that process visual information in more comprehensive and abstract ways.[53] V1 has many small neurons that represent the "pixels" of a visual image, while later brain areas, with lower neuron density, but perhaps more interconnectedness, analyze features of vision such as motion, borders, or faces. While the cerebellum has much higher neuron densities than the cerebral cortex, it also has mainly small cells. Average glucose use by a neuron, a measure of its metabolic activity, is about 10 times higher in the cerebral cortex than in the cerebellum.[54] Thus, the human cerebral cortex, with 16 billion neurons, is more than twice as expensive in terms of metabolic energy than the cerebellum with 69 billion neurons. Higher energy consumption has generally been an indicator of increased synaptic signaling between neurons.[55] Eighty-two percent of human brain mass is in the cerebral cortex, and higher levels of abstract thinking still appear to be localized in the forebrain of the cortex rather than the cerebellum.[56]

Neurobiologist Georg Striedter suggests that measurements of relative brain size, such as EQ, may be the best way to compare distantly related taxonomic groups, such as birds with mammals.[57] They allow comparison of animals of very different sizes and focus on species that have increased in brain size compared to an average animal of that particular body size. In closely related species, such as among corvids or apes, absolute brain size may be the best measure, with actual neuron counts being better than brain volume. In both relative and absolute brain measures, larger brains appear to have more modules, with the possibility of taking higher abstractions of data as information is passed on from one module to the next.

Genes and Brain Size

So far I have concentrated on brain circuits as the major determinant of brain function, but an additional level of activity is gene action. Recent research has found a number of genes that influence brain size, including microcephalin, ASPM, and HAR1. Major defects in these genes can cause newborns to have severely reduced brain size—as much as 70 percent smaller than normal.[58] All three genes show rapid evolution in the primate lineage, with ASPM and HAR1 changing the most since humans and chimpanzees diverged. It seemed at first that researchers had found the genetic basis for the nearly threefold increase in EQ between humans and chimps. But a study of a large sample of people in the normal range of IQ found no correlation between variants of two of these genes and intelligence.[59] It appears more likely that anything as complex as enlarging a brain involves the concerted action of a large number of genes, each making a small contribution to the outcome. A comparison of thousands of genes in humans and chimps found 169 genes that are expressed differently in the human brain than they are in our primate cousin, with 90 percent of them more active in the human brain.[60] A similar comparison of heart and liver genes between the two species found no consistent difference. So the human brain is not only larger, it is doing different things, and future years should shed light on what these are at the most detailed levels.

In the next chapter, I will extend the discussion of abstraction to a seminal period of human culture when new ways of tool making, art, and trade became widespread among our ancestors. Some of the earliest forms of religion also trace back to this period, and religion can be considered another kind of abstraction, for it posits the idea that the personality of an animal or person can be separated from its body and live in a realm separate from the physical world. A late-maturing brain, with extra areas for forming higher abstractions, may have fostered such concepts, which may have had important effects on the social organization of early modern humans.

8

The Brain and Belief

There was a "Big Bang" of human culture about 40,000 years ago in Ice Age Europe while our species still lived a hunter-gatherer way of life. This burst of creativity included widespread adoption of new techniques in material and mental culture. It was signaled by a new way of producing stone tools, but also introduced other advances such as heated shelters, eyed needles for sewing, widespread trade, and the use of new materials, such as bone and antler.[1] This revolutionary period also produced new kinds of art and religious practices, including elaborate burials with grave goods, the first known appearance of musical instruments, and widespread representational art like that found in the magnificent cave paintings of southern France and Spain.

We now know that this cultural explosion had a long fuse, dating back to hominid groups in Africa and evidenced in cruder stone blades at 280,000 years B.P. (before the present), special coloring agents like red ocher for decoration as far back as 900,000 B.P., and a bone harpoon point and pierced shells for body ornamentation dating from 80,000 years ago.[2] But what appears piecemeal in isolated early groups becomes a widespread cultural phenomenon during the Upper Paleolithic revolution, beginning about 40,000 years ago. More than 99 percent of known Pleistocene art dates from this period.[3]

Figure 8.1 presents one of the earliest known carvings, dating from 33,000 to 30,000 B.P. in southern Germany.[4] Made from a mammoth tusk, the carving is of a lion-man. Together with anthropomorphic figures found in cave paintings from around this period, this carving may give insights into the unusual frame of mind that became widespread

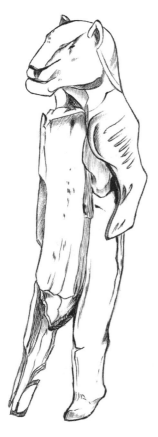

Figure 8.1 Ivory statuette of a lion-man.
One of the oldest known human carvings, about 28 centimeters (11 inches) high. (Original in the Ulmer Museum, Ulm, Germany)

during this period. The mixing and matching of disparate elements, with a lion's head put on the body of a human, may illustrate an unusual "cognitive fluidity" that became prevalent at this time.

Archaeologist Steven Mithen suggests that earlier groups of humans had multiple intelligences, each specialized for important aspects of survival. "Social intelligence" facilitated the complex interactions of group living, "natural history intelligence" allowed them to understand the behavior of other species in the wild, and "technical intelligence" fostered early tool making. But this was a "Swiss-army-knife" mentality, with separate brain areas devoted to different specialized skills. The modern mind, suggests Mithen, began with the blending of input from these different areas, joining elements into novel combinations.[5]

The human brain has an especially large number of modules for abstracting information. The prefrontal cortex (PFC), as discussed in chapter 7, is implicated in higher orders of abstraction, and it has an unusual amount of white matter to interconnect areas of the PFC with themselves and with more posterior areas of the brain. Humans are also unusual in the length of time they use to mature these connections, continuing the insulation process (myelination) into early adulthood. Experience—and not just genes—has an important influence on this "hardwiring." Such a brain is predisposed for the cognitive fluidity that Mithen proposes. It can abstract qualities from the raw data of the senses and mix and match them in new ways to find useful combinations.

The burst of innovation in the Upper Paleolithic period that fostered new forms of tools, clothing, ornamentation, and trade relations may have come from the kind of creativity represented by the figure of the lion-man—that is, the ability to combine ideas from different domains into novel combinations. As the spirit of an animal and a man can be abstracted from their bodies and combined in new ways, so qualities of materials like stone, wood, and antlers might be combined in new ways to produce new tools. As noted in the last chapter, analogical thinking employs prefrontal areas, and perhaps these too came to a new level of maturity at this time.

Representational art is actually a kind of abstraction.[6] Modern people are surrounded by images daily and forget what it would be like if you had never seen a drawing of anything. Two-dimensional and three-dimensional art relies on the abstracting of essential contours—it never depicts all the details of its model. When people first began to make pictures and sculptures, there had to be decisions about which lines best represented the nature of the real thing.

Contact with Another Realm

Another abstraction that became prominent during the last Ice Age is apparent belief in an afterlife—that is, that the soul can leave the body and live on in another realm. This is exemplified by a burial site of two adolescents in Russia dating from about 32,000 years B.P. The body of what apparently is a boy was buried with about 5,000 beads, a decorated cap, a belt with 250 fox teeth, ivory carvings, and an ivory lance. Nearby a girl was buried with more than 5,000 beads and other objects.[7] Before this period, few, if any, grave goods are found with burials. Our own species buried their dead at least as far back as 100,000 years B.P., and

Neanderthals from at least 70,000 years B.P., but implements are rarely found with the bodies.[8] From the Upper Paleolithic onward, grave goods become a regular cultural phenomenon.[9]

We cannot, of course, be certain that these practices signal belief in an afterlife, but studies of other hunter-gatherer societies that produce rock art indicate that their worldview is permeated with spirituality: both humans and animals have spirits that can leave their bodies, and their natures can blend. Such beliefs persist in groups that have pursued a hunter-gatherer way of life into modern times, such as the San of South Africa, a variety of Native American tribes, and some Australian aborigines. Similar depictions of animals and humans are found on rock faces in all these cultures, and in all cases, it turns out to be a religious art, more like the ceiling of the Sistine Chapel than a naturalistic catalogue of herds of game animals passing nearby.

Among the San of modern South Africa, the rock face is believed to be a veil over the spirit world. Paint, often mixed with the animal's own blood, has the power to coax the spirit of the animal from the netherworld, especially if it is its image that is painted. Some large animals, such as the powerful eland, can impart their strength to humans.[10] One can, in effect, become possessed by the spirit of an animal and gain some of its powers, transforming into an amalgam of animal and human natures. San rock paintings show people in various stages of transformation into animals, much like those found in cave paintings and sculpture from the Paleolithic. More than 50 cave paintings from the Paleolithic show beings that are part human and part animal. In general, animal images are much more frequent than human ones, reflecting perhaps the belief that their spirits were more powerful than those of humans.[11]

Large caves, such as the famous one at Lascaux in southern France, have wide chambers where a number of people could gather and witness a stately parade of animals painted on the ceiling. But there were also small side tunnels, barely large enough for a single person, where jumbles of drawings are sometimes found, one image drawn on top of another. Some researchers have argued persuasively[12] that these were chambers for sensory deprivation, where an individual could focus on spiritual experiences, and that the drawings are actually records of experiences of animal spirits coming up out of the rock. A jumble of images, one atop another, can be found there, probably because each of them represented a separate divine event, an encounter between human and animal spirit. Significant experiences were preserved, even though they overlay each other. These ancient people were consummate artists and could depict separate animals in large herds with excellent perspective when they wanted to, as shown in the large chambers. When one picture is overlaid on another it is not for want of understanding how to depict separate animals; rather it is likely recording one profound experience after another, like separate snapshots of memorable events. Further evidence that the rock surface was a membrane over the spirit world comes from Paleolithic drawings where an animal is partially emerging from the rock itself, such as a horse's head drawn on an outcropping of stone, implying that the rest of the animal is still behind in the depths of the rock.[13] Other paintings incorporate contours of the rock—for instance, a stone nodule may form the eye of an animal.

In hunter-gatherer society, the mediator between the human world and the spiritual realms is often a shaman. He has the ability to enter into altered states of consciousness and travel to the spiritual world, where he gains the power of special spirit-animals, which can then be used for healing the sick, to aid in the hunt, or to control the weather. Various techniques are used to attain these altered states, including music, chanting, dance, fasting, and psychoactive drugs. In San society, half the men and a third of the women can act as shamans, and everyone participates in the communal dance that invokes the spirit world.[14] This is in keeping with the egalitarian nature of hunter-gatherer societies, where everyone partakes of the fruits of the hunt and material goods generally are shared in common. Among the Tungus of the Arctic (the word *shaman* is derived from their language), a man or a woman can be a shaman. Communal meetings of the clan involve rhythmic music and dancing until the shaman falls into a trance and becomes a vessel for the spirit that they are trying to contact. Possession is a way of taming the spirits and inclining their favor to people, and no clan feels secure without its shaman.[15]

Shamanism may be the earliest form of religion and possibly the source of many traditions that come later. The idea of a tiered cosmos that has both physical and spiritual levels is one of the most enduring qualities of religion. The worshipper can bridge these realms and be infused, in varying degrees, with divine powers. The shaman is the first religious specialist, and he seeks to integrate the human and spiritual realms. The guardian animals can be seen as the forerunners of animal spirits found in totems and the animal-headed gods of various ancient religions, such as that of the Egyptians. Monotheism formed a higher, more universal sublimation of ideas about the divine, but retained in prophecy the tradition of religious possession.

When the shaman seeks to integrate the human and spiritual realms, a sacrifice is often necessary to achieve that harmony. In some cultures, the shaman is believed to die symbolically as his soul leaves his body to travel to the spirit world, where he becomes transformed, wholly or in part, into a spirit-animal. While in trance, the shaman's body often lies prostrate on the ground in front of the worshippers. Shamans sometimes danced in the skins of their guardian animal, and in many Native American tribes, every young man was expected to go on a vision quest, involving fasting and privation, until he had been contacted by his guardian spirit-animal.[16] In some societies, epileptics were preferred to be chosen as shamans.[17] The theme of sacrifice or privation as a way to reach the divine endures today in traditions like fasting, and the idea of a wounded healer who dies to help his people is, of course, a major theme in Christianity. Even the dome of a cave, with its controlled, filtered light, can be seen as part of the structure of churches, synagogues, and mosques, which often have a domelike shape in the central worship area. Although biblical religion goes back some 3,500 years, shamanism is at least 10 times older and probably held sway until the Neolithic period, some 10,000 years ago, when the hunter-gatherer way of life was gradually supplanted by settled farming and the beginnings of the first cities. The memories of a great flood that are found in many ancient traditions may hark back to the end of the last Ice Age, when retreating glaciers left behind huge bodies of water that ultimately broke through their embankments and flooded the land. Dim memories of these catastrophes persist in the oral histories of the first civilizations that came afterward. The hunter-gatherer way of life stretches back at least 10 times as long as what we call recorded history, and it may still influence our values—and perhaps even our genes.

Religion and the Maturing Brain

Our species, *Homo sapiens*, appeared about 200,000 years B.P. in Africa.[18] We migrated out of Africa through the Middle East about 100,000 years ago, arriving in eastern Europe about 45,000 years B.P. and in western

Europe at 40,000 years B.P.[19] For some 55,000 years in the Middle East, we lived side by side with Neanderthals, a species with a brain size at least as large as our own. There also does not seem to be any increase in the cranial capacity of our species from the time of our earliest appearance in Africa. Over that long period,[20] *Homo sapiens* and Neanderthals manufactured essentially the same tool kits, and their culture changed little. Only with the Upper Paleolithic revolution did the use of a variety of new tools and weapons become widespread within our species, and Neanderthals became extinct soon afterward, at about 27,000 years B.P.[21]

What fostered this creativity if it was not a change in brain size? Perhaps it was an emphasis on particular kinds of circuits in the brain that are good at abstracting and combining information. Religion may have introduced the first higher-order abstractions with the ideas that the spirit of an animal or human can leave the body and dwell in a realm separate from the physical world. It may thus have fostered abilities in the brain that were latent until cultural events evoked them. We saw something like this in language training with chimpanzees and dolphins that were capable of learning syntax. Their communication in the wild does not appear to have this level of complexity, and unique events appear to have evoked latent capabilities.

The frontal areas of the brain that are specialized for handling concepts and doing analogical thinking may have been influenced by religious practices to form the kinds of circuits that are good at blending disparate concepts, the kind of "cognitive fluidity" that Mithen finds flourishing in the Upper Paleolithic. We know that children were brought into the painted caves of this period from hand marks and footprints left there. More recent cultures with rock paintings and shamanism incorporate puberty ceremonies for both males and females.[22] Among the Quinault of the North American west coast, for example, boys made rock paintings of the visions they saw. Among tribes in California, shamans supervised rituals in which adolescents ingested hallucinogens and painted images on rock. In a variety of Native American cultures, young men were expected to go off on a vision quest to find their guardian animal-spirit.[23] The importance of influencing young minds persists today in traditions like the bar mitzvah in Judaism and confirmation in Christianity, and it is probably the source of the Jesuit maxim, "Give me a child until he is seven, and I will give you the man."[24]

Today we send our children to school from an early age to learn the symbols of language, math, and science. But in a period before any schools, the highly emotional ceremonies of entry into adulthood may have fostered the maturation of brain circuits adept at abstraction and symbolism.[25] Religion may have introduced the first higher-order abstractions, and these habits of mind were then applied to other areas of human activity. This may be the reason we find spirituality, art, and cultural innovation all flowering at about the same time in ancient history.

It is often suggested that language is the main reason for cultural advances in our species.[26] By having words to represent things, we can manipulate them in new creative ways. Perhaps language matured to the point where we could symbolically represent many things in nature and thus combine them in novel ways. Development of language probably played a role in the creative burst of the Upper Paleolithic, although because it is an activity that leaves no fossils, we probably will never know for sure.

But humans form higher abstractions in a variety of areas, many of them nonverbal.[27] Math, music, painting, language, and dance are processed by different combinations of circuits in the brain. The abstractions of classical music are as highly derived from sound as any language, yet a Beethoven symphony does not need any words to convey its meaning. Similarly, painting and ballet can reach very high levels of complexity, and words are scarcely able to explain why they move us. These activities do not all have to pass through language centers of the brain in order to achieve high levels of abstraction. There are savants of math and music who remain severely retarded in language. Psychologist Nicholas Humphrey shows drawings of an autistic girl who, between the ages of three and six years old, made beautiful line drawings of animals comparable to those found in some cave paintings, but whose vocabulary comprised only 10 words, which she rarely spoke.[28] Nor is sophisticated language necessarily needed to pass on ancient tool-making culture. In an interesting experiment, two groups of undergraduates were shown how to make Neanderthal-style stone tools, technical skills that are not easily mastered by modern humans. One group was given elaborate verbal instructions and a demonstration, the other only a silent demonstration. Both groups learned equally well, at the same speed (although there were some people in both groups who never could master the techniques).[29]

I suggest that humans are not capable of symbolism because they have language; rather, they have language because, as animals of high relative brain size, they have extra circuits that can achieve higher levels of abstraction in a variety of areas. A large-brained animal has more to express, and it can do this in sound or other modalities, just as it manipulates its environment in intricate ways with whichever appendage has evolved to the greatest dexterity.

Civilization and Its Discontents

With the melting of the ice, people domesticated plants and animals and were able to produce the first food surpluses, thus supporting the growth of the first villages and cities. But traces of the old hunter-gatherer way of life can still be found in the civilizations that followed. For people emerging from an Ice Age, a warm, steady sun must have been an especially welcome sight, and it is perhaps not surprising that they believed its continued presence was a fragile thing that they somehow had to foster. In ancient Egyptian civilization, as well as in Mesoamerica, the highest gods were identified with the sun, and human activities were necessary to insure its continued passage across the sky. As agricultural civilizations, they would also, of course, notice the importance of the sun in nourishing their crops.

Many of the gods of early civilizations were human-animal amalgams, such as the Egyptians' Horus, who had the body of a man and the head of a falcon. He was identified with the sun. The famous Sphinx has the body of a lion and the head of a human. These gods are similar to the combinations we saw in cave art but are a further kind of abstraction. In hunter-gatherer religion, a shaman could pass to the spiritual world and temporarily gain the power of spirit-animals. But now animal-human combinations have become permanent eternal beings before whom one bows in worship.

The reification of concepts that takes places in spirituality mirrors a differentiation that took place in social life. Egyptian society began in Neolithic times with small communities that settled along the Nile to farm the rich soil left behind by the annual Nile floods. Each village likely had its own local gods and chieftain.[30] These gradually coalesced into Upper and Lower Kingdoms that were finally united under Menes about 3100 B.C.E., in what is now called *Dynasty 1*. In Neolithic times, burials were simply in pits with the body laid on its side, although in Upper Egypt, grave goods are sometimes found with the deceased, indicating that people believed the spirit went on after death. The bodies might also wear amulets with pictures of animals, such as the gazelle, serpent, or lion, presumably in the belief that these imparted special powers to help the wearer in the afterlife.[31] Although the chieftain might have a somewhat larger grave, there were not yet significant differences in the status of people. The deceased were also not yet embalmed, because the hot desert sand that covered the graves quickly desiccated the flesh and left some resemblance to the body in life. This probably led to the idea that a person could persist in the afterlife as both a spirit and a physical person.

By the time of the unification of the two kingdoms, clear differences between classes of people had arisen. The rich had large tombs with underground chambers chiseled out of rock, and there might be a staircase leading down to these chambers from the surface. Above was a brick superstructure resembling a house with many rooms. Elaborate grave goods were supplied to the wealthy, including furniture, jewelry, cosmetics, and weapons, along with many jars of food. It is clear the Egyptians expected the dead to continue in many ways as they had in life, placing in the tomb food for a physical body that would come back to life. The poor had much simpler graves with few or no grave goods.[32]

The most elaborate burial was for the Pharaoh, who wore the double crown of the two united kingdoms. He was no longer simply human but the earthly embodiment of a god, the hawk-deity Horus.[33] In each dynasty, the heir to the throne was presented as the result of a divine union between the current king's wife and the chief god of the time (who would descend temporarily in the guise of her husband, thus settling the question of who really was in bed with the queen). The resulting god-king owned the whole land and its people by reason of his exalted state, but he was not free to simply be a despot. He too was subject to the divine laws of the gods and had to uphold order and justice on Earth.[34] The continued journey of the sun across the sky could be threatened, and without proper justice and worship on Earth, the universe might turn to chaos.[35]

The pyramids that were built above the underground burial chambers and that increased in grandeur with subsequent dynasties allowed the Pharaoh to travel from the underworld to join his father, the sungod, in the solar boat that sailed across the sky each day. The ordinary citizens of ancient Egypt were participants in this cosmic drama. Most of the peasants who built the pyramids were not slaves, but rather labored on the tomb during the three months when the Nile flooded and farming was not possible. They received food and shelter from the state in return for their work. This both maintained employment and gave ordinary Egyptians the chance to contribute in maintaining the divinely decreed order of things.

Thus, one of the first empires in history was built around a religious idea, as were nearly all early civilizations. Egyptian theology can be seen as an elaboration of several of the spiritual ideas of earlier hunter-gatherer religion: (1) The spirit of a person lived on after death and might use implements from life, such as weapons or jewelry. This is seen in the burials of both hunter-gatherers and early Egyptians. (2) Underground was a gateway to a spiritual realm. This seems obvious for people who buried their dead and believed in an afterlife. But this idea goes further, populating an underworld with a variety of spirits that humans can contact. The underworld is not the hell of Christian theology, for most of these spirits were benign—either the animal spirits of the hunter-gatherers, or gods like Osiris who helped Egyptians in the afterlife. The cave was an entryway into this realm for Paleolithic hunters, and an underground room, plain and unadorned, cut into the rock beneath the pyramid, was the place where an Egyptian left his coffin and entered the spiritual world. (3) Animal-human amalgams populated the spiritual realms. For hunter-gatherers these were probably temporary associations, such as the shaman taking on the power of a lion spirit and bringing this back to the physical world to help his people in some way. For later civilizations such as the ancient Egyptians, animal-human combinations became the gods themselves, and prayer was a way of invoking their special powers. The personality of the animal was a clue to the nature of the god. Horus, for example, one of the forms of the sun god, had a falcon's head because the flight of the falcon, with its freedom and power, seemed akin to the passage of the sun overhead.[36]

The reification of the gods paralleled the stratification of Egyptian society. The egalitarian nature of hunter-gatherer society, where food and other goods are generally held in common, was replaced by increasing class differences as Egyptian society became wealthier. This was held together by the religion of a god-king, so that in serving the ruler, an ordinary citizen was helping to uphold the divine order of the Egyptian cosmos. This ideology was amazingly effective, creating an empire that lasted, despite various upheavals, for thousands of years. It succeeded, in part, because of its syncretism, the ability to recognize the many gods of early Nile villages and incorporate them into a common pantheon, even as the separate settlements coalesced into a united kingdom. But this system also succeeded because it overcame the fragmenting nature of hunter-gatherer society.

Tribal Life

We do not know the nature of relations between the earliest settlements along the Nile, but a study of small villages in present-day cultures that are near the hunter-gatherer level of organization gives some clues. A global survey of 31 hunter-gatherer societies found that 64 percent engaged in warfare every two years and only 10 percent had peaceful relations with their neighbors.[37] The death rate for young males in tribal warfare is actually higher than for men during the two world wars of the twentieth century.[38] A study of intertribal warfare in New Guinea found adult male mortality rates of about 26 percent for a variety of tribes.[39]

The effect of intertribal warfare can be seen among the Yanomamö of the Amazon in South America, who have been intensively studied as one of the few remaining populations that have had little contact with modern life and remain near the hunter-gatherer stage of organization. Villages typically number about 90 people, many related to each other through male descent. They need to farm only about three hours a day, and the rest of their calories are supplied by hunting animals in the jungle. A village can grow, but when it reaches about 300 members, it usually fissions, and a group stakes out a new area in the jungle to build huts. Over time, suspicion builds between these encampments, even though there are kinship ties between them. Then the raids begin. A group of men from one village sneaks up on another village and may only fire arrows into the encampment before fleeing back to their own homes. But if they find a man alone outside his village, often they will fall upon him and viciously kill him. A woman found unprotected will be abducted, probably raped by the raiders, and then given as a wife to one of the men in their home village. Incessant warfare goes back and forth between Yanomamö villages, and it is estimated that 30 percent of their men will die violent deaths.[40]

Any system that can overcome the centrifugal nature of human suspiciousness and wield people into larger units would have powerful survival value. One of the effects of religion is to teach people that there are powers greater than themselves. It creates spiritual brothers and sisters under the purview of a father god. It leaps beyond mere genetic selfishness that looks out for just oneself and kin and can create larger communities of people.[41] The syncretism of early Egyptian religion gathered in one town after another and had powerful social consequences. The ability of the Egyptian state to summon large numbers of workers in the service of their god-king was also reflected in the size of the armies it could recruit to defend and enlarge the state.

There came a stage in early human evolution when we had mastered fire and improved our weapons to the point where the greatest remaining danger in nature was other human beings. From then on, being able to form larger human groups would have a distinct advantage for survival. Indeed, we may all be the descendents of people who could experience religious emotion and form larger cohesive societies because people who stayed in smaller units simply disappeared. Darwin recognized something like this when he wrote, "There can be no doubt that a tribe including many members who, from possessing in a high degree the spirit of patriotism, fidelity, obedience, courage, and sympathy, were always ready to aid one another, and to sacrifice themselves for the common good would be victorious over most other tribes; and this would be natural selection."[42] A recent mathematical study of ancient and modern hunter-gatherer societies shows that lethal group conflicts could have supported the spread of quite costly forms of altruism. That is, groups that were more cooperative had a high likelihood of prevailing in conflicts with groups that were less cooperative.[43]

Music and Dance

In the painted caves of Ice Age Europe, there may have been religious ceremonies that merged people together into something larger than the family unit. The first known musical instruments are found there, bone flutes dating from about 36,000 years B.P. These were not simple instruments and were probably reed voiced: the pipe was inserted into the player's mouth, which held a vibrating reed. One in France had four holes offset from each other and beveled to allow subtle finger control.[44]

Music is an important part of religious services to this day because it has the power to bring people together in a common rhythm, which may include singing, clapping, and dancing. Add to this the worship of a common deity and there are powerful forces to bind people together at both an emotional and intellectual level. Large united groups would have an advantage over smaller groups in many social activities, including hunting and warfare.

Music is a part of worship in many primitive societies. These are not short services reserved for the end of the week, but marathon sessions that may go on all night and involve the whole community. Their goal is to achieve both social solidarity and contact with the divine.

Science writer Nicholas Wade describes three societies near the hunter-gatherer level of organization where music and dance play important roles in their religious ceremonies. Genetic studies indicate that these isolated groups may be very ancient, not far removed from a human migration that left Africa some 50,000 years ago to populate other parts of the world.[45] The first is the Andaman Islanders, who probably arrived at their islands in the Indian Ocean during the last Ice Age, when sea levels were around 61 meters (about 200 feet) lower than today. Studies near the turn of the century, a time when they were still not greatly influenced by modern civilization, show that their dances started at night and lasted at least five to six hours. All able-bodied adults participated, with women singing sacred songs while the men danced. Wade quotes an anthropologist who visited the tribes from 1906 to 1908: "As the dancer loses himself in the dance, as he becomes absorbed in the unified community, he reaches a state of elation in which he feels himself filled with energy or force immensely beyond his ordinary state, and so finds himself able to perform prodigies of exertion."[46] These rituals were performed periodically to renew community solidarity and also before a battle, to fortify warriors from one village in their fight against men from another village.

The second group is the Australian aborigines, whom modern genetic studies show are all descended from a single stock that arrived about 45,000 years ago in Australia and remained isolated until the arrival of the Europeans. They too emphasized song and dance as part of worship in ceremonies that went on long into the night and might continue for days. In a reconciliation ritual, parties that had a long-standing quarrel would have a symbolic duel with blazing sticks, after which the dispute was never to be discussed again.

The third group is the !Kung of the Kalahari desert in Africa (! represents a clicking sound made in their language). The main religious ceremony is an all-night dance held around a fire underneath the stars.

The women sit in a circle singing and clapping, while the men dance around them. After a couple of hours of strenuous dancing, men begin to go into trance. They feel in contact with the spiritual world and may even feel their own spirit leaving the body. Physical healings take place, and the ceremony banishes forces of evil. The community is acting as a unit, drawn together by the rhythms of music and dance, mutually promoting the spiritual welfare of the participants.

Wade notes that the sacred songs of the !Kung are sometimes merely strings of vowels and nonsense syllables. Modern Pentecostal churches, which also strongly emphasize music as an integral part of worship, have singing and "speaking in tongues" during intense spiritual experiences. Although they claim these are ancient languages, they are strings of syllables with no known meaning. Wade suggests that worship, music, and dance may be older than language.[47] They arose at a very early stage of humanity when the need to form social groups larger than the family was essential for survival. They bound people together at an emotional level, and being able to work as a unit was crucial in both the hunt and in competition with other groups of foragers. Even today, music, dance, and worship have the ability to move us at subliminal levels that are hard to put into words. The earliest protoreligions may trace back to a time when there was still only protolanguage.

Group Selection

Until recently, most biologists doubted that selection for groups could take place in nature. It seemed that genetic selfishness, where the individual works to benefit himself and his kin, must always be stronger than altruism, where the individual sacrifices self-interest for the good of a group. The classic case of the latter involves honeybees, where workers will sting to defend the hive, and as a result of losing the stinger, die. These suicides seem like the ultimate self-sacrifice, but in the strange genetics of the order Hymenoptera—to which bees, ants, and wasps belong—workers are unusually highly related to their sister workers in the hive. In most animals, a parent shares 50 percent of his or her genes with a child. (The mother and the father each contribute half to a child's genetic makeup and are therefore 50 percent related to each child. By a similar logic, brothers and sisters on average share 50 percent of their genes.) In the Hymenoptera, males have only one set of chromosomes whereas females have two, with the result that the daughters of a queen are 75 percent related to each other. When they sacrifice their lives to defend the hive, they are protecting closely related kin. Living female workers are also sacrificing their reproduction. They have functional ovaries, although their activity is usually dormant.

Recent studies have shown that in some bee species, queens mate with up to 10 different males, so that workers in such hives are, on average, only about 30 percent related to each other.[48] So kinship may not be the whole explanation for *eusociality*, the term used for social species that use reproductive suppression. Termites comprise some 3,000 species and can have a caste system as stratified as the bees, and yet their genetics are like those of most animals, where parents are only 50 percent related to their young.[49] A study of a primitive termite family (Termopsidae) is instructive for issues of group selection. Unrelated colonies may inhabit the same log, and if they meet, they can merge. Workers retain the ability to differentiate into different castes, and should the reigning king and queen die, workers can become reproductives. In 73 percent of merged colonies, reproductives come from both colonies.[50] Large, merged colonies have advantages in survivorship and in battles with smaller neighboring colonies. Moreover, any pair that left a colony and tried to found a new one would have low chances of success.

The main issue appears to be not the degree of relatedness, but what a group is capable of doing. If a pair has only a 2 percent chance of survival by themselves but a 90 percent chance if they stay in the colony, even with only a 5 percent chance of becoming reproductives in that colony it is still in their genetic interest to stay with the colony ($0.9 \times .05 = .045$, or 4.5 percent). The odds change even further when a colony can build a structure that no two individuals could possibly build by themselves, like a large termite mound that is both air-conditioned and has a rock-hard protective cover.

There also seems to be a point of no return in colony complexity. Once workers differentiate to a certain level of specialization, the colony becomes a stable developmental unit that never reverts to simpler levels of organization. The unit of selection has changed, and one colony is competing with other colonies for resources.[51] Darwin believed natural selection could take place at multiple levels. Certainly individual and kin selection are important, but the effectiveness of one group in competition with another group will also affect the survival of individuals within that group. Ants have become so successful that the greatest danger in nature for an ant is another ant. Complex social life is such an effective strategy that although ants and termites constitute only about 2 percent of the some 900,000 insect species, they make up more than half of the total insect biomass on the planet.[52] These facts can be extrapolated to human beings and how we now dominate our world. Ants create large, complex societies through kinship relations, but humans have found other ways to get individuals to cooperate. According to Wade, "That is why ants don't need religion but people do."[53]

The Meaning of Sacrifice

The Bible recognizes the tension between belief systems and reproductive advantage. Abraham, considered the father of the three monotheisms, is told by God to sacrifice his only son, Isaac, on a mountain. He is viewed as a model of faith because he obeys God's command without question. As his arm is raised to slay his son Isaac, an angel stops him and God says, "[B]ecause you have not refused me your own beloved son, I will shower blessings on you and make your descendents as numerous as the stars of heaven and the grains of sand on the seashore" (Gen. 22:16–17).[54] In the terms of modern biology, he is being told that because he is willing to give up genetic self-interest, his gene pool will prosper. This strange logic makes sense in the context of group selection. A belief system that induces individuals to subordinate their selfish interests can create larger groups that will be victorious over other groups. The Bible understands that the most primal self-interest is reproductive, and it even inscribes this covenant on the penis. After promising Abraham that he will be the father of nations, God tells him to have all the males in his group circumcised: "My covenant must be marked in your flesh" (Gen. 17:13).

In terms of genetic self-interest, what is more dear to a man than his son and his own penis? (Woody Allen remarked that his head was his second favorite organ.) Yet this is precisely what the Bible demands be subordinated to a higher principle. It is part of a theme of sacrifice that is almost universal in religions worldwide. Jews were expected to make regular donations of animals and grain to the temple in Jerusalem, and the sacrifice of goods, animals, and even humans is a pervasive aspect of primitive religion. Even religions that do not believe in gods have sacrifice. The Kwaio of the Solomon Islands worship their ancestors, who crave pork and demand sacrifices of pigs. Illness or misfortune can befall the living if the spirits are not appeased.[55] In Christianity, Jesus himself has become the sacrifice that reconciles sinful humanity with God.

It is a strange notion, really, that a spiritual being should want food. It was clear that the god or spirit was not really eating the food. The Bible says cooked food gives off a smell pleasing to God (Num. 15:1–16), and it was understood that the Levitical priests were due a portion of the meat and grain because they did not inherit property the way other Israelites did (Deut. 18). The Kwaio say the ancestor spirits like the smell of cooked meat and actually consume only the soul of the animal. The cooked body itself is divided among the worshippers in a "spirit of unconditional sharing." In the Muslim annual sacrifice of the ram, meat is shared among families, and gifts are given to the poor.[56]

In terms of group cohesiveness, these rituals make sense. Individuals are taught that there are higher principles than self-interest and that they must give up some of their goods to support the group. They will reap the benefits later in terms of group support or defense but must learn to put off immediate self-gratification. The problem is to find a principle great enough to induce people to overcome their native selfishness. An all-pervasive spirit that controls health and good fortune and also has the ability to read peoples' minds is impressive enough to induce people to subordinate themselves. Wade also points out that primitive societies lack a court system and police, and therefore "fear of divine retribution keeps almost everyone in line with the prevailing rules and moral code."[57]

Anyone who doubts the ability of belief systems to create powerful groups should contemplate the birth of Islam.[58] Before Muhammad began preaching his message in 613 C.E. (Common Era), the Arabian Peninsula was populated with scattered Bedouin tribes that had no organization above the tribal level. No individual could hope to survive without the protection of his group, so a murder between different tribes, for example, always had to be avenged by a reciprocal killing.

Muhammad's monotheism taught a new identity. *Muslim* literally means "one who submits," and *Islam* means "submission to God." When the traditional tribes objected to what Muhammad was teaching his followers, a unique event happened on the Arabian Peninsula. For the first time, men took up the sword against their own relatives in defense of the prophet's cause. Muhammad eventually defeated the polytheist Meccans, and then welcomed them as converts into the fold. Within 100 years after his death, the Muslim Empire spread from Spain to the borders of China.

Anyone who thinks it is merely material factors that determine history should contemplate this development. Before 630 C.E., the scattered Bedouin tribes had done nothing of world-historical importance, and they lived in a desert environment with few ecological advantages. Within a hundred years of adopting this new faith, they were engaged in the most advanced science and architecture in the world and had the most extensive empire. Trade flowed between Europe and Asia; a common religion created new economic associations, not the other way around. It is those glory days of the Islamic empire that the Muslim fundamentalists hope to recreate, not realizing that Muhammad was revolutionary and progressive for his time, not one who returned to the old ways.

When Muslims pray, they bend and touch their foreheads to the ground, symbolizing submission to a higher power. A common gesture of prayer in most religions is two arms raised skyward, and this is also the universal military sign of surrender. A faith that inspires people to subordinate their immediate self-interest for the good of the group has the power to create large societies that prevail over smaller groups of people. Both Christianity and Islam followed the implications of belief in one universal God by seeking adherents worldwide. Judaism, the original monotheism, has not, for most of its history, actively sought converts. Its fate among the larger monotheisms illustrates the dangers of remaining too small a community. As Wilson points out, religions tend to be charitable only to members of their own group, and the strength of in-group solidarity can be matched with the fierceness of out-group hostility.[59]

In-groups and Out-groups

Lethal raiding and the abduction of females is something we share in common with our close living relatives, the chimpanzees. Anthropologist Richard Wrangham and Dale Peterson have done an extensive study of "apes and the origins of human violence" and find this combination rare out of 4,000 mammal species and 10 million other animal species. In part this is due to similarities of social organization and ecology. Both we and chimpanzees live in patrilineal, male-bonded communities where females regularly move on to other communities to find mates, thereby minimizing inbreeding.[60] We are also both high-EQ animals that supplement our diets with meat to supply the high caloric requirements of large brains.

Recall from chapter 5 an account from Jane Goodall's study site: when a chimpanzee troop split into two neighboring territories, a kind of warfare developed between troops that is rarely found elsewhere in animal behavior. The larger group conducted raiding parties into the neighbor's territory, and if they found a lone male or female, they viciously attacked. Their victim usually died soon afterward from wounds. The attackers used tactics usually reserved for prey, including breaking bones, tearing off flesh, even drinking blood from wounds.[61] After eliminating all the males and older females one by one in a nearby small community, the larger community took over its territory and added the remaining young females to its own. Similar warlike behavior has been documented between chimpanzee communities at three different study sites in Africa.[62]

Most surprising, the rivals at Goodall's site had once been allies while they were still members of the same troop. But the declaration of separate territories, under the leadership of different dominant males, signaled a complete change in social relations. These tactics make sense from a selfish, genetic point of view. By appropriating territory for themselves, the winners gain greater resources, and by taking females from their rivals, they have more chances of passing on their genes to the next generation. Yet despite this simple, brutal logic, few species go to the lengths of chimpanzees and humans in warring with rivals.

Perhaps the ability to abstract and form categories contributes to this unusual behavior. A rival group is seen as an obstacle to be eliminated rather than as fellow beings. Goodall sees a parallel with the worst aspects of human conflict that separates in-groups and out-groups and that justifies the most brutal tactics toward members that no longer belong. In humans, this often begins with a period of dehumanizing the enemy, proving that they are not beings like ourselves and so they deserve to be eradicated. A high-EQ mind that is able to apply categories may classify individuals as part of an out-group and be so strongly led by conceptual behavior that it overrides any feelings of kinship. The high objectivity that allows sophisticated modification of tools may also form complex strategies for eliminating rivals. Whereas allies are highly prized, embraced, and their wounds cared for, enemies are met with unusual violence. Wrangham and Peterson describe such an enrichment of emotional responses, which has been one of the major themes of this book: "All the great apes, we know, have especially sophisticated brains. . . . They therefore amplify the range of tactics that individuals can use to interact with and manipulate each other. Some of these tactics are affectionate, and some are violent. . . . Intelligence turns affection into love and aggression into punishment and control. . . . The intense violence of apes arises partly from the very elaboration of their cognitive abilities."[63]

9

Energy Flows

Consider the old millstream. A waterwheel was built over a stream with buckets to catch falling water. The weight of the water turned the wheel, and this water was dumped from each bucket before it returned to its starting position so that the cycle could begin again. The revolving wheel performed useful work, such as turning millstones to grind corn. From this simple example comes an important principle: energy gradients can be used to perform useful work. The gradient in this case is water falling in a gravitational field, but there are many other energy gradients in nature that can be harnessed to perform work. The Earth exists in an energy gradient between a hot sun and cold outer space. The energy of light is used by plants to build up organic tissues. Specifically, they use light energy to boost electrons to higher energy levels and then store them in high-energy bonds in molecules like sugars. These electrons can fall back "downhill" to lower energy levels, much as water can fall downhill in a gravitational gradient. When we burn tissues like cellulose (which is made of the sugar glucose) in wood, electrons are falling down to lower energy levels as they combine with oxygen. Fire, of course, can be used to perform useful work—provided it is contained in a way that does not burn down the house. When we metabolize food for energy, such as in consuming sugars, we are letting electrons fall to lower energy levels in gradual steps, but the overall reaction is the same as burning wood. It is the difference between a gentle fall in multiple steps and a raging fire, which is much more difficult to contain.

You cannot build a functioning waterwheel over a lake. There has to be a gradient. Even though the lake is high up on a mountain, unless water has somewhere to fall to, it cannot be harnessed to do work. This applies to energy in all its forms—chemical, nuclear, or mechanical; there has to be a gradient, a flow of energy, before useful work can be performed.

Also, the transfer of energy will never be 100 percent efficient. All the energy of falling water will not be captured in grinding corn. Some is lost to the friction of turning wheels. All the energy of burning gasoline does not go into turning the tires of your car. In fact, the losses are quite high. The best gasoline engines are only about 25 percent efficient. The rest of the energy of combustion goes into heat, friction, and noise. From this comes a second great principle of energetics: in any energy exchange, some of the energy will be lost to disorder. In fact, this is a general principle of nature that goes by the term *entropy*. Entropy is what we call a general tendency in nature for things to go toward disorder. Take a room, for example. You clean it up, and you notice over a fairly short period that it tends to return to a disheveled state in which you have trouble finding the things you want. There are a few ways to have order, but a million ways to have a mess, and so over time the most probable configurations take place. Ultimately, entropy is a statistical law. It states that the most numerous arrangements are likely to happen over time. At this level, it seems obvious, but there are important consequences.

Because disorder is more probable than order, it provides a direction to the arrow of time. If I open a perfume bottle in a large room, the odor will gradually waft around the room. Before I opened the bottle, I had a kind of order: all the perfume molecules were confined to the bottle, with the unscented air in the rest of the room. But after opening the top, the perfume molecules begin to infiltrate the air, and vice versa. Over time, the most probable configurations take place. The perfume molecules are no longer in just one place, but are likely to be everywhere, and the same is true for the air molecules; that is, the air and the perfume are about equally mixed.

If I show you a movie of a man smoking a pipe, and the smoke gathers out of the air, condenses into a little cloud, and then disappears down his pipe, you know the movie is being shown backward. Clouds of smoke disperse, they do not spontaneously condense (although storm clouds can be another matter, which we will consider shortly). On the other hand, if I showed you a movie of two pool balls hitting each other and then ricocheting away, you would have a hard time telling if the movie were showing forward or backward (I am ignoring here the slight slowing of the balls that occurs because of friction). Newton's laws of motion are invariant to time. Every action has an equal and opposite reaction, and this applies whether balls are moving forward or backward. But entropy provides an arrow to time; it predicts the direction of an ensemble of objects.

Entropy can be measured and is an important part of chemical reactions. When glucose is metabolized with oxygen to release energy, a complex six-sided molecule of sugar is broken down to six simpler carbon dioxide (CO_2) molecules and six water molecules. Ninety-eight percent of the energy released comes from electrons in glucose falling to lower energy levels as they are handed over to atoms that bind them more tightly. But 2 percent of the energy comes from an increase in entropy as a large complex molecule of sugar is broken down into simpler molecules.[1]

The second law of thermodynamics states that in every energy transformation, some of the energy will be lost to disorder. When we metabolize sugar for energy, such as during exercise at the gym, the by-products (carbon dioxide and water) are smaller than the original sugar molecules, and we also give off heat. Heat is considered the lowest form of energy, for it radiates in all directions and cannot be recovered. The first law of thermodynamics states that energy cannot be created or destroyed, it can only be changed in form. The second law states that there is always a cost to doing business; in every energy exchange, you will have less useful energy at the end than you had at the beginning because some has leaked away to randomness. If you could trace down every bit of energy, at the end you would see that it is equal to the amount you had at the beginning, but its distribution has changed; more of it exists as low-grade heat than at the beginning.

It is said that eventually the universe will suffer a heat death. All differences will even out, all objects will be at the same very low temperature (slightly above absolute zero), and no work or new order will be possible in the universe. Yet here we are, some 14 billion years after the Big Bang, and everything is still proceeding with furious activity. Where does the original difference, the original gradient, come from that powers all this activity of star burning, planet formation, and life? It comes from the Big Bang itself, from the expansion of the universe that is opposed by the universal gravity between all things.[2] A giant rubber band has been stretched, representing the gravitational attraction between all matter in the universe. Because the original fireball had some lumpiness to it, it is contracting in various places into galaxies and clouds and giving up the gravitational potential energy that was put into it by the original explosion. Gravitational energy, in turn, ignites fusion inside stars, providing sunlight to plants and, in turn, food for animals. In each energy exchange, some of the energy is lost to low-grade heat, so that we will eventually dissipate the huge bank account that was provided by the original expansion of the universe. But that day, when all things have evened out to one low-grade temperature, is still very far off.

Dissipative Structures

A big puzzle is why order happens at all if things generally drift toward greater disorder. There is a natural tendency for complex things to decay, for metals to rust, for gases to mix, and for water to flow downhill. But put a rock in a stream, and you may see some of the water double back on itself. It takes some of the downward energy of the stream to curl over and move uphill for a while, until it falls again into the general downward motion of its neighbors and heads downstream again. But viewed from a distance, water seems to be doing the impossible: it curls back on itself and flows uphill for a while.

There is a whole class of structures like this in nature that seem to defy the second law of thermodynamics. If I have a box with a chamber of cold air and a chamber of hot air and I open the partition between them, the air will quickly blend. The hot, fast-moving particles will encounter the slower ones, and they will even out to one average, intermediate temperature. The study of such systems is known as *equilibrium thermodynamics*. They take place in closed boxes where conditions are allowed to reach equilibrium.

But when a cold front comes down from the north and it encounters warm air from the south, the two systems often do not simply blend. Frequently, storms develop at their interface, and they might even grow into hurricanes or tornadoes. These structures bear some resemblance to life. They start small, mature to full form, and then gradually die away. They feed off the energy gradient between hot and cold air masses, and they organize billions of molecules into coherent flows. They are termed *dissipative structures* because they actually help dissipate the energy difference between the two fronts. The study of such systems is known as *nonequilibrium thermodynamics*. They are not closed off from their sur-

(a)

(b)

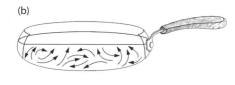

(c)

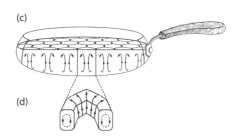

(d)

Figure 9.1 Creating Bénard cells.
A pan of oil is heated from below and its molecules go through increasingly chaotic motion in (*a*) and (*b*) as they absorb heat. In (*c*) they change abruptly from conduction to the regular flow of convection, forming hexagonal Bénard cells throughout. The flow pattern of a single cell is shown in cutaway in (*d*). (From Eric Chaisson, *Cosmic Evolution: The Rise of Complexity in Nature* [Cambridge, Mass.: Harvard University Press, 2001], 60. Adapted by permission of the author)

roundings but instead depend on a continual flow of matter and energy through them.

To understand how such structures work, consider a more limited example known as *Bénard cells.* These can be created by heating a pan of silicone oil from below and keeping a uniformly cool layer of air over the top of the oil. At first the oil acts like any familiar liquid being heated on a stove. In figure 9.1*a*, molecules of liquid are moving faster as they absorb heat, and they distribute energy by bumping into one another at random, a process known as *conduction*. In figure 9.1*b*, random motions are increasing with greater heat. But at a critical temperature (figure 9.1*c*), the oil molecules suddenly shift from conduction to convection. They form hexagonal cells all through the pan; the random motions of heat have been replaced by cells with definite borders within which the oil molecules are moving in a highly organized fashion. The details of one cell are shown in figure 9.1*d*. Heated oil is expanding and rising in the center of each cell, cooling at the top, and then sliding down the outer edges of the hexagon to repeat the cycle when it reaches the heated bottom.[3]

Bénard cells illustrate several principles about dissipative structures in general. They all exist at energy gradients and actually help dissipate the difference between two energy states. Rather than working against the second law of thermodynamics, they facilitate it by increasing entropy production. The difference between the heated bottom and cold top of the fluid is partly dissipated by the random motions of the oil molecules. But when they shift from random conduction to organized convection, they actually speed up the mixing of hot and cold. They are therefore dissipating a difference and increasing the entropy of the overall system. Herein lies the paradox of these structures. A kind of order is being created in the organized flow of oil molecules inside the hexagons. But that order is helping to break down a larger order: the difference between the hot and cold layers at the top and bottom of the pan. The increase in order of the hexagons is less than the increase in disorder by dissipating the overall energy gradient, so net disorder increases and the second law of thermodynamics is preserved. The second law does not specify the pathway that must be taken, so long as overall entropy increases. It does not care how you pay your taxes as long as the check arrives.

Another characteristic of dissipative structures is they often have points of decision, bifurcation points. The oil is heating with ever more random motion until it suddenly goes through a kind of phase change from chaos to hexagonal form. Studies show that these hexagonal structures arise within a limited set of energy gradients. Too shallow a gradient, and you get only random mixing. Too steep a gradient will overpower and destroy the convection cells. But so long as energy is provided at the right intensity, the cells will persist, feeding as it were, off the energy flow.

There are obvious parallels with metabolism and life, although I do not mean to suggest these cells are alive. Plants and animals live within the energy gradient of sunlight, and so long as it remains within a limited temperature range, the complexity of life can persist by feeding on an energy flow. The organized structures of life, like the hexagonal cells, are counter to the universal tendency of nature to disorder, and they can maintain themselves only by continually importing energy. For animals, this process is called *eating*.

Flames, storms, river eddies, and the vortex that drains water from a bathtub are other examples of dissipative structures found widely in nature. They appear to have a steady form only because a constant traffic of molecules is passing through them. They persist only so long as they have matter and energy to feed on. Storms in general and hurricanes in particular are maintained by the energy gradient between large masses of hot and cold air. A hurricane, for example, forms in a temperature differential between two layers, as in a Bénard cell, but in this case it is between a warm sea below and a cold upper atmosphere. Air molecules rise upward through the center of the storm cell, creating dramatically lower pressures in the "eye" of the storm. The organized system of a hurricane may be a thousand miles across, and it can persist as long as it gains power from an energy differential between warm sea and cold upper air. Once it passes over land and loses that power source, it quickly dies away.[4] Hurricanes do not form at every temperature differential between the ocean and upper atmosphere. It takes just the right gradient to start these self-perpetuating structures. Such conditions are provided in the Canary Islands, off the coast of Western Africa. In the autumn, winds blowing past the mountains on the Canaries can form swirling eddies of moist air. With the right temperature differential and sufficient moisture, these vortices can mature into hurricanes that travel on the trade winds to the United States.[5] In some seasons, a virtual assembly line of these storms can be seen marching across the Atlantic.

A hurricane, like a Bénard cell, actually serves to dissipate the gradient above and below it. Evolutionary theorists Eric Schneider and Dorion Sagan give a telling example of this process in miniature, called *tornado in a bottle*. Turn a large soda bottle upside down and watch the liquid as it goes "glug-glug-glug" out the neck of the bottle. Billions of small molecules are trying to exit the narrow bottleneck and they do so in a disorderly fashion. Now fill the bottle again and give it a couple of swirls before releasing the mouth. The liquid, turning in a vortex, exits the bottle smoothly in a much shorter time.[6] In just such a fashion, an organized form can help dissipate a gradient, or, if we reverse the logic, a gradient can lead to the organization of form. In neither case do we violate the second law of thermodynamics because the form helps dissipate a larger kind of order: the energy differential between two larger masses. It can be said that "nature abhors a gradient." Just as entropy seeks to even out the difference between all the perfume molecules in a bottle and the air in the rest of the room, it seeks to even out the temperature difference between two masses.

The form that is created represents a kind of order, but its existence is temporary. It persists only so long as it can tap into the energy of the gradient, and the amount of order it creates must always be less than the total movement toward disorder. Like a rock in a stream that causes some of the water behind it to curl backward and flow "unlawfully" uphill for a while, that motion is always less than the total mass of water flowing downhill, and it will eventually rejoin the general downward movement. In a similar way, life on Earth exists in the great stream of energy that flows from the sun. It uses a very small percent of that energy to build up the ordered structures of life. But it too is fated to decay eventually and rejoin the general downward stream toward disorder.

Physicist Ilya Prigogine pioneered research on dissipative structures and won a Nobel Prize for his work in 1977. One of his findings was that these structures can have multiple bifurcation points,[7] as shown in figure 9.2. In figure 9.2a, we see increasing amounts of energy applied to the system over time. At critical points, such as Ec1, it takes on a new structure as a stable state. We saw this in the hexagonal Bénard cells that suddenly appear when heating silicone oil.

Some dissipative structures have multiple stable points. As the energy applied to the system varies, it shifts to different configurations as the most efficient way to degrade the gradient applied across it. The pathway chosen may have a history. In figure 9.2a, because one stable state was chosen at energy level Ec1, another pathway becomes possible at Ec2, and another at Ec3. Small fluctuations may incline the system to one side of the bifurcation or another.[8]

There are clear parallels to changes in biological systems, as shown in figure 9.2b. As the environment changes (E1 to E3), it imposes new pressures on a species that may speciate into new forms. Some of these forms will be bound for extinction (such as B), whereas others may be the ancestors to future species (C). Patterns like this are found in the theory of punctuated equilibria, which states that a particular species persists for a long time until it meets some environmental challenge that causes it to split, in a relatively short time, into two or more species. Some of these descendents will go extinct whereas others will persist for a long time and become the progenitors of future species. A pattern similar to this is found in the evolutionary history of horses.[9] All life forms can be seen as dissipative structures that live off energy gradients available to them as food, and they may be forced to speciate when the environment changes enough that they must find a new way of making a living. Every species, of course, has a history that influences the possibilities available to it at the next bifurcation point.

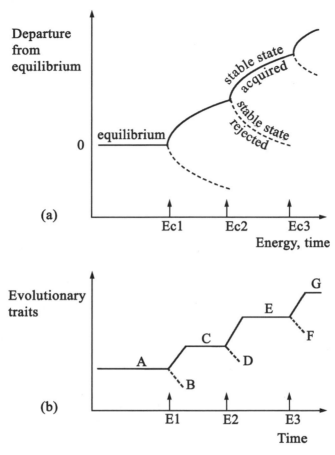

Figure 9.2 Dissipative structures in physics and biology.
These structures can reach bifurcation points at which they shift to new dynamic steady states. Small variations can incline the system to one pathway rather than another: (*a*) changes in energy applied to a dissipative structure cause it to change state; (*b*) environmental changes cause a population to speciate, leading to a species that survives (as in C) or goes extinct (as in B). (From Eric Chaisson, *Cosmic Evolution: The Rise of Complexity in Nature* [Cambridge, Mass.: Harvard University Press, 2001], 55. Adapted by permission of the author)

Energy and Complexity

Order, complexity, and *information* are related terms. A more complex structure will have more order to it than a simpler structure of the same mass. It will take more information to describe a complex structure than a simpler one. I define complexity as having more parts and more functional relationships between those parts. By these criteria, a car is more complex than a bicycle. The term *functional* is important because it might be said that a wrecked car has more parts and arrangements between those parts than an intact car. But in a wreck, many of those arrangements will no longer be functional, and so we can say it has less order and less complexity. It may take more information to describe the scattered parts of a wreck, but that does not make it truly more complex than an intact car. Only the functioning relationships between parts count in a measurement of complexity.[10] We might measure complexity in terms of the minimum amount of information, in bits, needed to describe a structure.[11]

Although we might be able to do this with the blueprint for a car or a bicycle, it is very difficult to measure the information content of objects in nature. How many parts and relationships shall we count for a storm, a cell, or the organs of a whole animal? Chapter 2 addressed this in biology by estimating complexity by the number of cell types in an animal body and information content in terms of the number of genes or relative brain size of a species. But it showed no simple relationship between body complexity and the number of genes.

There may be another way to estimate complexity in terms of energy use. In general, it takes more energy to build or maintain a complex structure than a simpler one. It will take more energy to assemble a car than a bicycle, and more power is needed to drive the more complex vehicle. We saw in dissipative structures that an energy gradient can help create an ordered form like a Bénard cell or a hurricane. The greater the gradient, the more complex the structure it might build.

Astrophysicist Eric Chaisson has proposed a functional measure of complexity in terms of the energy used by a system. Complexity is measured by the amount of energy passing through a structure per unit time per unit mass, in ergs/sec/gram. Much as temperature gives an average of the speed of countless molecules that we cannot hope to measure individually, Chaisson's metric of *energy rate density* proposes to measure the complexity of a system by the amount of energy needed to maintain its structure.

Table 9.1 shows his calculation of energy rate densities for a variety of objects, and figure 9.3 shows their emergence in relation to time, with details for different stages at the right. Each stage provides a setting for the evolution of the next. Thus, galaxies provide an environment in which

Table 9.1. Some Estimated Free Energy Rate Densities

System	Average Energy Rate Density (erg/sec/gram)	Approximate Age (billions of years)
Milky Way	0.5	12
Sun	2	5
Earth's geosphere	75	4
Plants generally	900	3
Animals generally	40,000	0.5
Society (modern culture)	500,000	0

SOURCE: Eric Chaisson, "Energy Rate Density as a Complexity Metric and Evolutionary Driver," *Complexity* 16, no. 3 (2011): 27–40.

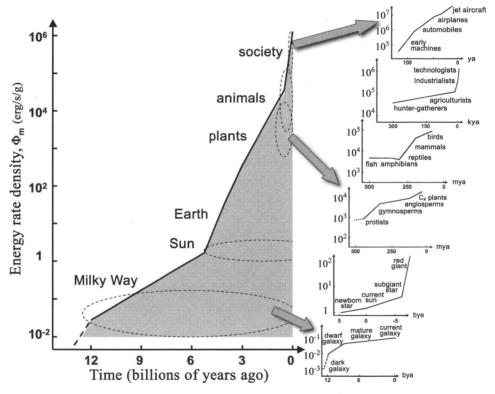

Figure 9.3 Chaisson's energy rate densities for a variety of open, nonequilibrium systems.

The rise in energy rate density over time as more complexly organized structures use higher energy rates. Details of specific areas provided on the right. bya = billions of years ago; mya = millions of years ago; kya = thousands of years ago; ya = years ago. (From E. J. Chaisson, in a chapter to appear in *The Self-Organizing Universe: Cosmology, Biology, and the Rise of Complexity*, ed. C. Lineweaver, P. Davies, and M. Ruse [New York: Cambridge University Press, 2012]. Adapted by permission of the author) star formation can take place, stars condense from a disk that may include a retinue of planets, and—in our solar system, at least—one planet has provided an environment for life. Chaisson's calculations show increasing rates of energy flow in this series. Note that although the total energy flowing through the sun is hugely greater than that flowing through an animal, the specific rate per unit mass is not.[12]

Chaisson uses our sun as an example of a star, noting that it was fainter and more massive in its youth. The sun has been heating up at the rate of about 1 percent each 100 million years and slowly losing mass through its solar wind and radiation. Currently it has a luminosity of 4×10^{33} erg/sec and a mass of 2×10^{33} grams, giving it an energy rate density of 2 erg/sec/gram. In its early days, when it was made almost entirely of hydrogen, it had about half this value. In about the last 5 percent of its lifetime, there will come a point at which helium will ignite in its core and the sun will swell into a red giant phase. As it fuses helium into carbon, it will form another inner layer to its stratified structure, and its energy rate density will increase about tenfold.[13]

More massive stars show even more clearly the relationship between structural complexity (accumulating shells of ever-denser elements) and the rate of energy flux through the star. Stars 10 times more massive than the sun fuse ever-heavier nuclei with corresponding increases in energy rate density (in erg/sec/gram): hydrogen (600), helium (1800), carbon (2600), and oxygen (4000).

Chaisson finds the Milky Way to be fairly typical of galaxies in its overall energy flux. He calculates its luminosity at all wavelengths and divides the amount by the Milky Way's mass, which includes its spiral arms and interstellar gas, arriving at an energy rate density of about 0.05 erg/sec/gram. He does not include dark matter, the nature of which is still enigmatic, but if it were included it would bring the value of energy flux even lower.[14] Current theory indicates that galaxies form by the merging of many early dwarf galaxies, which likely had lower energy flux and less complexity than mature galaxies. Chaisson proposes that galaxies are relatively simple structures defined by just six parameters (e.g., mass, size, spin, and age), but these are correlated with each other so they may be summarized by just one variable: present-day mass.

For plants, he finds major changes in energy flux between single-celled algae and phytoplankton, gymnosperms, and angiosperms. In animal life, the major distinction is between ectotherms (cold-blooded animals) and endotherms (warm-blooded animals). Chaisson extends his metric to stages of human civilization and the power of human machines. Further details on these various systems are provided in a section that follows. One of the most attractive qualities of his theory is that it characterizes physical systems over 20 orders of magnitude and nearly as great a reach of time, using identical units of measurement at all levels of complexity.

Energy Use and Life

The idea that increasing energy flows sustain increasing complexity of structure accords well with the patterns of organic life discussed in chapter 2. I noted there that the more complex structure of the eukaryotic cell is supported by roughly 1,900 times more power per gene than the average prokaryotic cell.[15] There are also grade shifts of increasing metabolic rates in going from unicellular eukaryotic organisms to cold-blooded vertebrates to warm-blooded vertebrates.[16]

The basal metabolism (energy used at rest) of single-celled organisms is less than that of multicelled organisms, and the basal metabolism of cold-blooded vertebrates (poikilotherms) is lower than that of warm-blooded species (homeotherms). These differences are plotted in figure 9.4, indicating grade shifts in the use of energy. These differences also correspond to changes in the information content of organisms in genes, brains, or both, and it suggests that more complex organisms may need more energy to run.

These trends continue with different groups of mammals. The primitive monotremes have a lower body temperature (30°C [86°F]) than marsupials (35°C [95°F]), and marsupials have a lower body temperature than placental mammals (about 38°C [100.4°F]), which are the last group of mammals to diversify. These differences in body temperature correspond to differences in basal metabolism. Marsupials generally have only 70 percent of the standard metabolism of placentals of the same body size.[17]

The brain is one of the most complex structures in the body and also one of the most expensive to run. In humans, it occupies only 2 percent of the body but accounts for about 20 percent of our basal metabolism. Its metabolic cost is eight to 10 times that of skeletal muscle per unit mass.[18] A recent study of 1,247 mammal species found a positive correlation

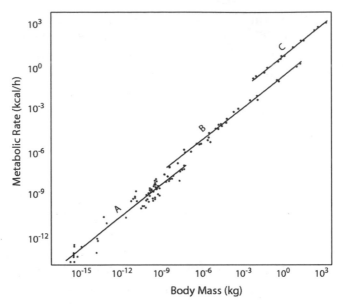

Figure 9.4 Differences in metabolic rates.
Metabolic rates for (A) unicellular organisms at 20°C (68°F), (B) animals dependent on their surroundings for body heat (poikilotherms) at 20°C (68°F), and (C) animals with internal body heat (homeotherms) at 39°C (102.2°F). Note that the scale signifies 1,000-fold changes (kcal/h = thousands of calories per hour). (From Knut Schmidt-Nielsen, *Animal Physiology: Adaptation and Environment*, 5th ed. [Cambridge: Cambridge University Press, 1997, after Hemmingsen, 1960], 195. Adapted by permission of the publisher)

between relative brain size and basal metabolism rate (BMR).[19] In young animals, the energy demands of the brain are even greater. In human babies, for example, the growing brain draws on 60 to 70 percent of available energy.[20]

The large study of mammals cited here removed the effects of both body size and phylogeny from its analysis. Body size must be corrected for because there is a well-known relationship between BMR and an animal's overall dimensions. A mouse has a faster metabolism than an elephant, and there is a regular relationship of decreasing metabolism for increasing body size for all animals in between. It takes more energy to run an ounce of mouse than an ounce of elephant. This relationship possibly relates to the geometry of surfaces and volumes and the need to exchange materials like nutrients and oxygen across surfaces. If we think of the body as a sphere (which more and more Americans approximate), volume increases as the cube of the radius, but the surface only increases as the square of the radius. Thus, as volumes get bigger, surface areas increase at a slower pace, and bodies must slow down to be adequately supplied. The reasons for a linear relationship over a large range of body sizes is still being debated.[21] But a proper study of the effect of brain size on metabolism must control for the effect of body size alone.

Modern studies of correlation in biology also correct for phylogeny. Species that have a common ancestor are more likely to share characteristics and not be truly independent data points. Modern studies use "independent contrasts" to correct for this. The 2009 study already cited supersedes an earlier study of a smaller sample of mammals that did not use independent contrasts and did not find a correlation between brain size and BMR.[22]

In birds, there is no correlation between BMR and brain size, but there is a positive relationship between brain mass and dietary energy content,[23] suggesting that a larger brain does have higher energy requirements.

Energy and Human Evolution

Increased energy needs were a vital factor in the long human voyage from smaller-brained ancestors. A study of 176 nonhuman primates found a positive correlation between brain size and both BMR and gestation period.[24] Primates in general have a larger relative brain size than do other mammals, and this applies as well to fetuses in the womb. Much of brain growth takes place shortly before and after birth in primates, so the mother's caloric intake likely affects what she can supply to the baby.[25] Large brain size correlates with both the time of incubation in the womb and energy turnover by the mother.

Humans are the primates with the highest encephalization quotient (EQ) of all—some three times that of our near living relative, the chimpanzee.[26] As babies, our brains use about 65 percent of the body's energy needs, and the young chimpanzee brain draws only about 10 percent less. Chimps mature about age 10, but humans have drawn out the life cycle so that adulthood and brain maturation take place later, and this probably affects our extended energy needs. Between the ages of 10 and

18, a chimpanzee is using about 11,000 calories (11 kilocalories [kcal]) per day, but a human is using about 42 kcal per day.[27] Both the pruning and insulating of the human brain go on into our early twenties.

Genetic changes also indicate increased metabolic needs for larger brains. Microarray studies can reveal large-scale changes in the expression of many genes at the same time. Various studies show that in human evolution, as opposed to the evolution of nonhuman primates, there has been an increase in the rate of gene expression changes in the brain, and these changes generally involved up-regulation (increased output) from these genes. By comparison, genes in the liver and heart show as many increases as decreases in output in humans as they do in nonhuman primates (i.e., chimpanzee, orangutan, and rhesus macaque, for comparison). Many of these genes in the brain are involved in neuron function and activity at the junctions between neurons (synapses). Other up-regulated genes in the brain are involved in the use of oxygen for metabolism.[28] These studies indicate that brain evolution has sped up in humans and that increased neural activity probably required more ample energy supplies to support it.[29]

Evolutionary changes in the lineage that led to modern humans also show a link between increasing brain size and rising energy needs. The australopithecines lived in Africa between 4 and 1.2 million years ago (mya) and had brain sizes anywhere from 430 to 530 cubic centimeters (cc), about one-third that of modern humans. *Homo habilis* lived between 1.9 and 1.6 mya, during which time brain size increased to 600 cc (36.6 cubic inches). Early *Homo erectus* lived roughly around the same time as *Homo habilis* and had a brain of about 850 cc. The *Homo* species possessed considerably smaller teeth and jaws than the earlier australopithecine species, as well as a reduced gut size (as judged from trunk proportions), indicating that they ate richer, higher-calorie foods that were easier to digest.[30] Low-grade foods like leaves and grass require longer guts to break them down.

Sites with human fossil remains show increasingly sophisticated tools from about 1.6 mya, including hand axes and cleavers.[31] Some researchers suggest that later *Homo erectus* mastered fire and that this further increased the calories available, especially from starches. Cooking boosts the digestibility of wheat by 34 percent and more than doubles that of tubers like potatoes. Surprisingly, it does not increase the food value of meat, although the smell of barbecue probably increases its allure and fire kills bacteria, reducing the chances of food-borne illness.[32]

Most large primates have long guts to digest low-quality foliage. Humans have much smaller guts and a reduced colon. The gastrointestinal tract is almost as expensive to operate in terms of calories as the brain. Some researchers have suggested there was a trade-off: humans paid for the high-energy needs of enlarged brains by shortening their guts, and they were able to do so by consuming higher-quality foods that were easier to digest.[33]

Overall, humans are also less muscular than other primates, but we carry more fat on our bodies, and our babies are especially chubby.[34] The alluring curves of females in breasts and buttocks are mainly fatty tissue (sorry, biology takes the romance out of many things). Fats serve for long-term storage of energy but are also important for myelinating (insulating) neurons in our brains. Overall, we live on high-quality foods that are easy to digest and we use fire to help break down tissues. A large, complex brain is one of the main reasons for our high-energy needs. In humans it accounts for 20 percent of our resting metabolic rate, whereas in other primates it is responsible for only 8 to 9 percent of their rate.[35]

As discussed in chapter 2, the increased complexity of life on Earth has correlated with increases in available oxygen, which made higher levels of metabolism possible. Oxygen accumulated slowly in the Earth's atmosphere, and simple bacterial cells can live at 1 percent of current oxygen levels but single-celled eukaryotes, with a much more complex internal structure, need at least 2 to 3 percent.[36] Multicellular organisms evolved at oxygen levels of about 5 percent, and rapid rises in oxygen levels are associated with other signal events in the history of life, including the conquest of the land by insects 410 mya and then by vertebrates about 350 mya, as well as the increase in placental mammal body size beginning 65 mya.[37] The enlarging brain of hominid species is also correlated with growing energy needs, as already described.

Thus increases in the complexity of cell structure (prokaryotes to eukaryotes), body plans (multicellular organisms with loose tissue organization versus complex organs), and brain structure (larger brain size in brain:body ratios) are associated with greater energy usage. It follows from the simple proposition that more complex things should require more energy to build and function than simpler things. Chaisson's calculations indicate that this principle applies to both living and nonliving forms, and nonequilibrium thermodynamics seems to bear this out. For if structure can arise as a way to dissipate energy gradients, more energy might support greater complexity of structure.

Energy and the Biosphere

Not only have bodies become more complex over evolutionary time, but so has the complexity of the biosphere overall. Figure 9.5 shows an exponential increase in family diversity since the dawn of multicellular life. This pattern holds true separately for vascular land plants, marine invertebrates, insects, vertebrates (tetrapods) on land, and life as a whole.[38] Despite major extinctions of species, such as by the asteroid that wiped

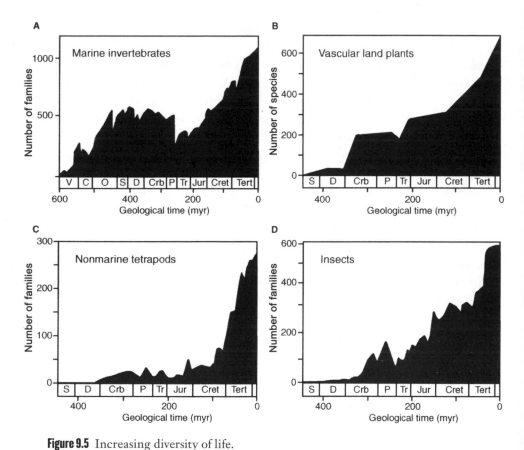

Figure 9.5 Increasing diversity of life.
The rising tide of diversity in the sea and on land in both the animal and plant kingdoms (myr = millions of years; o = present time). (From M. J. Benton and B. C. Emerson, "How Did Life Become So Diverse? The Dynamics of Diversification According to the Fossil Record and Molecular Phylogenetics," *Palaeontology* 50 [2007]: 26. Adapted by permission of Wiley-Blackwell, Oxford, U.K.)

out the dinosaurs, life has always managed to bounce back and resume the march to greater diversity. The worst extinction known, at the end of the Permean Era when the supercontinent Pangea formed (the P to Tr transition on the graphs in figure 9.5), was only a temporary interruption of this relentless trend. (It should be noted that families are usually composed of many species. If only one species survives, the family survives, and it may expand again later into many new species. Thus, there can be high extinction rates for species, but it does not show up as drastically at the family level.)

Paleontologist Geerat Vermeij describes a general increase in the metabolism of life in both animal and plant communities as they grew in diversity over evolutionary time. Land plants became more productive: the ancient tropical forests of the Carboniferous period (359 to 299 mya) were made of giant club mosses (lycophytes) that were probably much slower growing than modern rain forests. These plants had low rates of water transport in a simpler circulation and high levels of decay-resisting compounds that are energetically expensive to produce, resulting in slow rates of turnover and low levels of herbivory.[39] Their accumulated debris formed vast coal deposits around the world that are still being mined. In the late Carboniferous these plants were replaced by more productive ferns, and this change is associated with a broad diversification of plant-eating animals. The evolution of fast-growing flowering plants (angiosperms) increased the rate of turnover of plant biomass and supported a wider diversity of herbivores. Angiosperm grasses invented a novel growth pattern, extending the leaf at its base so that the grasses could survive cropping at the ends, and thereby made possible vast herds of herbivores on grasslands. Vermeij describes increasing cycles of productivity from plants to animals to decomposers like microbes and fungi, "all contributing to land vegetations in which nutrients move rapidly and through many organisms in a self-made microclimate that is particularly amenable to plant growth and consumer sustenance."[40]

In the sea, also, there were trends to greater diversity and productivity. Phytoplankton are small marine organisms that account for about 50 percent of all the photosynthesis on the planet, and they sit at the base of marine food chains. Their species diversity and productivity have increased over time, as measured in sediment layer thickness. In strata from the early Paleozoic era, about 40 species have been found. This number rose to about 400 by the end of that era, 250 mya. In the Mesozoic era, phytoplankton diversified to some 1,000 species, and

6,440 species are counted today.[41] This broadening of the base of the food chain supported increasing biomass and activity of animals at higher feeding levels. Both brachiopods and mollusks form hinged shells with muscles inside to snap the shells shut. Branchiopods were dominant in the Paleozoic, but they have much less visceral tissue and lower oxygen needs than the mollusk species that succeeded them. Similar changes in the amount of soft tissue can be seen in the transition from ancient trilobites to later crabs and lobsters.[42]

Rates of predation apparently have risen over evolutionary time as evidenced by the percent of shells with bored holes, or crushed and repaired parts, found in succeeding strata under the ocean.[43] In fish, there are long-term trends toward stronger bites and greater jaw flexibility. Early fish species of the Paleozoic had their upper and lower jaws rigidly attached to the skull, so the mouth could do little more than open and close. By the Mesozoic, a reorganization of the muscles and bones allowed greater flexibility, letting the fish expand its mouth cavity and allow water and prey to be sucked in. Advanced teleost fish, which greatly diversified in the late Mesozoic and early Cenozoic era, have even greater jaw mobility, with the upper jaw able to protrude past the lower, allowing a variety of specialized feeding strategies.[44]

Broadening of the base of the food chain, in either plants on land or plankton in the sea, made possible an increasing variety of lifestyles at higher feeding levels. With more active predation, this led to faster cycling of energy through the system as a whole. Paleontologist Richard Bambach and colleagues show that an increase in species diversity in the ocean led to a speedup of species interactions over time.[45] In a theoretical study, they defined a wide variety of potential ways of life using six possible feeding strategies from passive filter feeding to active predation, six levels of feeding from ocean surface to the bottom, and six speeds of motion. Combinations of these three dimensions defined 216 possible modes of life. They then asked how many of these ecospace possibilities were filled during different periods. At the first appearance of multicellular fossils, about 575 mya, 12 modes of life were present, most of which were likely suspension feeders, and only four appear to be animals that moved regularly. By the mid-Cambrian period, this number nearly tripled to 30 modes of life, and from that period to modern times, it tripled again to about 90. This development roughly follows the increased richness of the fossil record at the genus level, which quadrupled during the early Paleozoic and then doubled again from that period to recent times. Over that long time span, both speed of movement and predation increased, and predator-prey arms races were likely responsible, in part, for the diversification of life. In the Paleozoic, predators averaged 15 percent of animal assemblages, by the Mesozoic they were 24 percent, and today they average 36 percent of global marine diversity.[46]

Paleontologist Sarda Sahney and colleagues found similar results in a study of vertebrates on land.[47] All terrestrial four-legged animals can be traced back to an ancestor that crept out of the water about 400 mya. Vertebrates today comprise some 30,000 species in more than 300 families. The researchers defined 16 possible diets, three body sizes, and six habitats combining into 288 potential modes of life on land. The fossil record indicates that ancient amphibians occupied about 10 of these possible ways of life, reptiles and dinosaurs increased this to about 30 modes, and the evolution of mammals and birds raised this to about 70 occupied modes of life.

The rate of diversification apparently sped up over evolutionary time. In the Paleozoic (up to 251 mya), amphibians maximally added one-half family to a mode of life per million years. In the Mesozoic (251 to 66 mya), reptiles added up to one family per million years, and in the Cenozoic (66 mya to the present), birds and mammals have added up to three families per million years. In the last 30 million years, mammals and birds have tripled the rate of diversification into different lifestyles.

Many biologists deny any long-term trends in evolution and see the history of life as just one thing after another. Ideas of progress have been banished from biology, but distinguished evolutionary theorist E. O. Wilson writes that "the overall average across the history of life has moved from the simple and few to the more complex and numerous. During the past billion years, animals as a whole evolved upward in body size, feeding and defensive techniques, brain and behavioral complexity, social organization, and precision of environmental control—in each case farther from the nonliving state than their simpler antecedents did. More precisely, the overall averages of these traits and their upper extremes went up. Progress, then, is a property of the evolution of life as a whole by almost any conceivable intuitive standard, including the acquisition of goals and intentions in the behavior of animals."[48]

Diversity and Energy

Vermeij suggests that the biosphere has been increasing in both diversity and power with time: "Successive dominants at every scale of organization become more powerful, more productive, and internally more diversified, and they increase both their reach of control and their independence from their environment."[49] Warm-blooded species like mammals and birds have six to seven times the speed and endurance of reptiles and amphibians of the same size. Although a reptile can manage a short burst of energy, such as when a crocodile lunges for its prey, it cannot pursue its quarry with the long-term speed and endurance of a warm-blooded animal. This capacity allows mammals and birds to forage more widely and be active in a wider range of climates and seasons.[50] This homeostasis comes at great cost. The energy needed to maintain a warm-blooded mammal is about 10 times that of a lizard or plant of the same weight. But the fitness advantages, in terms of scope of activities and finding food, outweigh the inefficiencies. Brian Keith McNab found that mammals with higher rates of metabolism also had higher rates of potential population increase, thereby increasing their fitness.[51]

Diversity begets diversity. New species make possible new alliances or new relationships between predators and prey. T. H. Huxley likened this process to filling a barrel with large stones. The spaces between stones made room for pebbles, and these in turn left room for sand. As an ecosystem fills with species, it creates microhabitats where other species can make a living. The many niches in a coral reef illustrate this principle. Also, the broader the base of primary production—either phytoplankton in the sea or plants on land—the more trophic levels can be supported above it. On average, there is only about a 10 percent feedthrough from one feeding level to the next. Because of the entropic losses of transferring energy from one level to the next, it takes about 1,000 kilograms (kg [2,205 pounds]) of plants to feed less than 100 kg (220 pounds) of grasshoppers, and 100 kg of grasshoppers can sustain about 10 kg (22 pounds) of mice. Thus food webs rarely extend beyond more than four or five trophic levels.[52]

The diversity of life appears to depend on the amount of energy available. A number of studies support a positive relationship between plant diversity and productivity. The most productive environments are the most species rich.[53] An experiment that manipulated the number of plant species in an area found that net primary production in grams per square meter per year increased with more species richness.[54] Food webs become more complex in going from the poles to the tropics, and this correlates with increasing amounts of sunlight available for plant production. There are many possible reasons for the great species diversity found in the tropics, but increased available energy is surely one of the main ones. Ecologist Lionel Johnson states: "As food webs are lengthened and become more complex, the number of energy transfers increases, the more work is done, and the greater is the energy flux."[55]

Paleontologist Stephen Jay Gould says the trend to increasing the diversity of life is simply diffusion from a single common ancestor and has no underlying cause other than randomness.[56] But the transitions to higher levels of complexity outweigh those going the other way. There are no known cases of eukaryotes producing prokaryotes, multicelled organisms reverting to single-celled species, or large-brained homeotherms reverting to small-brained ectotherms.[57] Certainly not all nature is striving toward greater complexity. We saw that r-selected, low-information organisms have distinct advantages in some competitions over K-selected, high-information organisms. We should expect simpler organisms to persist along with more complex ones, as they certainly do.

Life as a Thermodynamic Flow

The unity of life on Earth suggests that it can be seen as a single thermodynamic system that has increased in breadth and power with time. All life probably diverged from a single common ancestor, as witnessed by the common genetic code used by all cells, both prokaryotic and eukaryotic. Over time, it has found new ways to degrade the energy gradient that flows from the sun past this planet, and that increased flux, as measured by Chaisson's energy rate density, supports increased complexities of structure. This is not some mystic élan vital; it follows from the nature of dissipative structures in nonequilibrium thermodynamics, and this principle applies to both organic and inorganic energy flows. Over time, both living and nonliving systems have developed as ways to dissipate the incoming solar flux.

Every species tends to expand its numbers to make maximal use of the energy available to it, and where there is a niche with unused energy, a species is likely to evolve to fill it. In this way, life as a whole resembles a thermodynamic flow that makes maximal use of the energy available to it.

By storing energy in its tissues, it may delay the ultimate degradation of that energy, but in using that surplus to reproduce, it maximizes the numbers of a population that can take up the highest form of available energy.

Both inorganic and organic processes are involved in dissipating energy. By absorbing and processing solar radiation, the Earth creates about 10 times more entropy than if sunlight were simply reflected off its cloud cover back into outer space.[58] Both the water cycle and the carbon cycle on the planet foster this process of degrading incoming solar energy. The average humidity on Earth is about 60 percent. A totally saturated atmosphere would have 100 percent humidity and nearly continuous clouds covering the planet, which would reflect much higher levels of sunlight back into space. Instead, we have continuously dynamic weather systems that absorb and process much more solar energy.

Rainfall weathers rock and puts calcium into the oceans, where it binds to carbon dioxide absorbed in water and forms limestone (calcium carbonate). It thereby takes an important greenhouse gas out of the atmosphere and lowers temperature on the planet.[59] Without this mechanism, the Earth would probably be a lot hotter. The water cycle alone, therefore, affects carbon flux on the planet, but this interaction, it turns out, is enhanced by the presence of life. The growth of plants increases rock weathering, and roots create a spongelike structure that holds water. A study of Hawaiian volcanic flows shows that lichens increase weathering of rock 10 to 100 times. More complex plants accelerate this process.[60] Plants also take in carbon dioxide to build their tissues, thereby further cooling the planet. Freezing helps break up rock, increasing the weathering cycle that leads to more binding of carbon dioxide in the oceans. Overall, the intertwined cycles of water, carbon, and life have created a thermodynamic system that takes in high-grade sunlight and radiates low-grade heat back into space. The complex structures built up in our weather systems and varied life forms are helping dissipate the energy gradient between our star and cold outer space and can be thought of as natural consequences of the laws of thermodynamics.

Ecological Succession

Thermodynamic patterns can also be found in a variety of ecological successions. Studies of different habitats—forests, grasslands, sand dunes, and marine shores—often find a similar pattern of species succession. The earliest colonizers of a new area are often rapid, r-selected growers of small size and fast life cycles. With time, species diversity grows along with increases in size, complexity, slower life cycles, and more hierarchical structure in the ecosystem as a whole.[61] An example of such succession is found in eastern forests of the United States, which begin as grasslands and are followed by shrubs, then pine forest, and they mature as tall stands of oak and hickory trees. More ways of life are possible in the mature ecosystem, with more kinds of predation, mutualism, parasitism, and so forth.

Both information content and homeostasis increase in such patterns of succession.[62] There are more species and more relationships between species, so it would take more information to describe the system. It is more homeostatic in that it has more ways to adjust to perturbations. The mature system is also a more efficient degrader of incoming energy. More incoming solar energy is converted to biomass, and it is retained in the ecosystem longer than in the shorter life cycles of the simpler system. Satellite data show that rain forests are cooler than grasslands or deserts because they are more efficient gradient reducers of energy.[63] In the Sahara, 41 percent of incoming solar energy is reradiated as long-wave heat, whereas in the Amazon only 17 percent is.

It is curious that only about 1 percent of sunlight is converted by photosynthesis into plant tissues. But Schneider and Sagan estimate that a huge 66 percent of incoming solar energy is used by trees to power their circulation in a process known as *transpiration*.[64] All plants must move water from roots to leaves, which for a tree may be over a distance of hundreds of feet. Like animals, plants have tiny channels that extend all over their bodies to carry fluids with nutrients to all their cells. But no one has ever found a beating heart inside a plant's body. How do they accomplish these feats of circulation? These tiny channels end in small openings at their leaves called *stomata*, and plants use solar power to evaporate water out from their stomata. Water is very cohesive, and as a tiny column of water evaporates out of an opening, it, in effect, pulls a whole thread of water behind it. Many threads together, evaporating out of many stomata, can add up to a lot of water. A typical oak tree will transpire some 10 kilograms (6 gallons) of water in a day, and a large rainforest tree as much as 1,180 kilograms (742 gallons) of water per day. An oak will use as much as 58 million calories of energy to move water through its many channels. (In hot, dry conditions, plants become stunted because they must close their stomata to conserve water. Circulation stops, and with it, growth.)

A mature forest is a prodigious degrader of energy, which is why forests are cooler than grasslands, deserts, or cities. Aerial surveys show lower temperatures in a forest than in a pond within the forest, both of which are cooler than a clear-cut section of the forest. Blacktop roads, as you would expect, reradiate a lot of heat, as do the downtowns of cities, and both are much hotter than surrounding rural farmland. Clear-cut forest areas have surface temperatures of about 50°C (122°F), whereas old-growth forests are at about 25°C (77°F), reflecting their much higher energy use for both growth and transpiration.[65]

Ecosystems tend to mature to the highest productivity they can within the constraints imposed on them, such as available sunlight, soil nutrients, and so on. Productivity is measured in the total organic matter produced by the animals and plants in the ecosystem, in calories, whether it is consumed or not. When ecosystems are disturbed, whether by man or natural disasters, they generally fall in productivity, indicating that they are at maximum productivity in normal circumstances.[66]

The Economy

Money can be thought of as a form of energy, and an economy as the structure of an energy flow. Money has the ability to command the energies of others by paying for their labor. In fact, its value can be measured in calories. A barrel of oil has 5.8 million British thermal units (BTUs) of energy, equivalent to 1.46 billion calories. At $75 per barrel, that is about 19.4 million calories per dollar. That seems cheap, but it does not take into account the costs of refining the oil, or the losses inherent in burning it in a machine.

Chaisson has shown the increasing use of energy at different stages of human civilization.[67] In terms of energy rate density, he calculates that (1) hunter-gatherers of a few million years ago used about 10^4 erg/sec/gram; (2) agriculturalists of several thousand years ago used roughly 10^5 erg/sec/gram; (3) industrialists of 200 years ago, 5×10^5; and (4) Western society today, about 10^6. The growth of the modern world's economy has paralleled closely the rise in energy consumption. Between 1900 and 2000, global energy use rose about seventeenfold and world economic product (adjusted for inflation) increased 16-fold.[68]

Increased energy consumption fueled the increasing power available to human society. The drafting of animals such as horses and oxen to aid in human labor augmented the power available by sixfold to eightfold.[69] Inventions such as the internal combustion engine multiplied this further and showed steady gains in power. Ford's model T delivered 15 kilowatts of power. By the 1980s, cars had outputs of 50 to 120 kilowatts.[70] Engines can also be rated by the amount of energy they deliver per gram of weight. The Wright brothers' four-cylinder engine delivered 0.1 watt per gram; B-29 bombers of WW II were capable of about 1.5 watts per gram. The Saturn 5 rocket that delivered men to the moon had a power:weight ratio of 1,000 watts per gram.

Increased energy input in agriculture has led to steady gains in output, which has supported expanding populations of people. Table 9.2 shows the intensification of food provision from early hunter-gatherers to modern agriculture in terms of energy per hectare (1 hectare equals about 2.5 acres). Since 1900, total crop harvests have risen about sixfold, but energy input to farming, in terms of fertilizers and mechanization, has risen about eightyfold; the trend clearly is unsustainable.[71]

As the available energy grew, economies became more globalized and integrated. World trade in 1945 amounted to 1 percent of the gross world product (world GDP).[72] By 1960, it was 24 percent, then 38 percent in 1985 and 52 percent in 2005. Thus, trade between countries has become more than half of all the world's economic activity (which is

Table 9.2. Intensification of Food Provision and the Population Densities It Is Able to Support

	Energy Value of Food Harvest (billion joules/hectare)	Density (people/km²)
Foraging	.003–.006	.01–.9
Pastoralism	.03–.05	0.8–2.7
Shifting agriculture	10–25	10–60
Traditional farming	10–35	100–950
Modern agriculture	29–100	800–2000

SOURCE: Adapted from Vaclav Smil, *General Energetics: Energy in the Biosphere and Civilization*, Environmental Science and Technology (New York: Wiley, 1991).

near $50 trillion in size).[73] In 2009, global trade was growing faster than world GDP.[74]

Communications and travel have fostered this process of global integration. In 1990, there were already millions of international tourists; by 2004, the number had increased a further 78 percent. By the early twenty-first century, about 100 million U.S. passengers were involved in 800,000 international flights each year. Americans studying internationally quadrupled between 1990 and 2007, and in many large U.S. universities 10 to 15 percent of the students are from other countries.[75]

The information content of the world has grown, both in the amount needed to describe its complexity and in the internal communication among the Earth's residents. The integration of the world is exemplified by Internet usage, which grew about 152 percent in North America between 2000 and 2011—and 1,037 percent in Latin America, 707 percent in Asia, and 2,527 percent in Africa. The largest total number of users in March 2011 was in Asia, at 922 million, with 476 million in Europe and 272 million in North America.[76]

The growth of social networks follows a typical J-shaped exponential curve. This is the shape of population increases and compound interest, where growth begets even greater growth at an ever-increasing rate. LinkedIn, a business-oriented networking site, took 16 months to gain its first million users. By February 2010, it boasted 58 million users, and gained its most recent 1 million in only 11 days. Facebook took almost five years to sign up 150 million users, but doubled that amount in the next eight months.[77] In the first quarter of 2011 it increased from 585 million to more than 665 million users.[78] It is perhaps inherent in the nature of information growth to have exponential patterns of increase. Businesses and researchers must know everything their competitors know and then do better.

The exponential increase in information was made possible by progress in devices that handle data. Between 1945 and 1999, the fastest computer speeds increased 10 million times (by seven orders of magnitude). From 1970 to 1995, the density of transistors that could be put on a single chip rose more than 10,000 times (Moore's law).[79] This concentrated information processing requires intense energy flows that result in high levels of entropy production, shed from the computer as heat. Modern microchips have to dissipate heat at a rate greater than 1 megawatt per square meter (roughly 10.8 square feet). This is more heat than the space shuttle encounters when reentering the Earth's atmosphere, and is just one order of magnitude less than the flux through the sun's photosphere.[80] One of the main challenges of modern computer design is to vent this excess heat, which has the potential to fry the circuits.

The growth of information technology fosters increased energy flow in business. Buyers are quickly put in touch with sellers and can assess local conditions. Individuals anywhere can buy and sell stock of many international companies at their computers. The trading of shares in efficient markets reduces the gradient between price and value.[81] Communications enhance the economic integration of the world. In 1914, there were about 3,000 international companies. By 1970 this number had reached 6,000; by 1988, 18,500; and by 2000, 63,000.[82] In the four years between 2003 and 2007, foreign investment by private equity funds in emerging markets went up about tenfold, from $3.5 billion to $35 billion.[83] There has been an accompanying growth in international nongovernment organizations (INGOs) concerned with philanthropy, human rights, and the environment. In 1900 there were 200 such organizations; in 1960, 2,000; in 1980, 4,000; and more have come into existence since. In the 1990s, Friends of the Earth had more than 700,000 international members, and Greenpeace claimed over 6 million members worldwide.[84]

Many of the themes we have considered so far are evident in the growth of human civilization: the relationship between energy, complexity, and information content. Increasing energy flows support increasing complexity of structure both in machines and the fabric of society. The system becomes more internally integrated as it grows in both size and power. Expanding international trade and communication facilitate this change. The information content of the system—both in the number of bits needed to describe the whole system and in the internal flow of information over its communication hubs on computers, satellites, and switchboards—has increased dramatically. The pace of life speeds up as changes in one area require adjustments in interrelated parts. The nearly daily updates now sent for computer software programs show this accelerating pace.

These are not just peculiarities of the growth of human civilization. There are parallels with what has happened to life over evolutionary time. The number of species, like the number of international companies, has increased as the system has grown in energy use. The increasing diversity of life made possible more complex food webs that had higher productivity and processed more incoming solar energy at different feeding levels. The information content of the natural world increased in the amount of information that would be needed to describe the growing diversity of life overall, in the number of genes needed to build more complex body plans, and subsequently in increased brain:body ratios. Mammals and birds increased nearly 10-fold in relative brain size compared to species that evolved before them, as described in chapter 3. Perhaps an increasingly complex environment required more complex brains to map it for animals that became dominant in their era.

High relative brain size is associated with more complex communication and social relationships in animals. Thus the organic world has become more interrelated with time. Flowering plants do not just rely on the wind to disperse pollen, but form specific relationships with animal pollinators. Mammals and birds have longer-term, more complex relationships with their young and with each other than did groups that evolved earlier.

What all these systems, living and nonliving, have in common is that they are energy flows. They have become more complex over time as measured by the information needed to describe their structure and the amount of internal communication between their parts. They have risen above the leveling effects of entropy by utilizing the power of an energy gradient, and the more complex they have become, the more energy they have needed to maintain their internal organization. They go through cycles of youth, maturity, and decay, and evolve to use more efficiently the energy gradient on which they feed. In this sense, the laws of thermodynamics reveal a unity between the organic and inorganic realms and how they develop over time.

10

The Origin of Life

If life is another form of thermodynamic flow, as the last chapter indicated, its occurrence may be widespread across the universe. Life as we know it makes use of the most abundant atoms produced by stellar burning, so it may be a special form of matter that self-organizes at energy gradients under the right conditions. As we saw, the principles of nonequilibrium thermodynamics are very similar for living and nonliving forms.

We cannot yet say whether life exists beyond this planet, but we have found that the precursors to the molecules of life are much more widespread and easily made than previously thought. This suggests that the evolutionary processes described so far may be prevalent all over the universe. In this chapter and the next, I wish to show that conditions for the increase in complexity and information content that we find on Earth may be widespread in the cosmos. This chapter will focus on the origin of life and the next will consider the prevalence of habitable planets where life might begin its evolutionary journey. There are at least three sources for molecules that could be precursors to life, and they have the potential to contribute to each other.

1. Vast clouds of dust and gas drift through interstellar space, composed mainly of primordial hydrogen and helium—the two most abundant atoms produced by the Big Bang—and enriched with heavier elements expelled by stars in the last stages of their lifetimes. The average temperature of these clouds is only a little above absolute zero, and yet a lot of chemistry is going on as the clouds encounter ultraviolet light

from stars they drift past and cosmic rays that are constantly crisscrossing the universe. More than 140 molecules have been identified in these clouds by the distinctive spectra they produce in microwaves that can be monitored on Earth.[1] Most are organic molecules with carbon chains containing up to 13 atoms. There are also significant amounts of methanol, formaldehyde, hydrogen cyanide, and carbon monoxide. Many of the building blocks of life can be formed from these molecules by simple chemistry. There are probably many more complex molecules forming in space, but their spectra cannot yet be resolved with ground-based instruments.

We now know that many of these molecules originate near the parent stars that produced the atoms that compose them. Stars live by fusion, burning primordial hydrogen and helium and gradually building up heavier elements in their cores. Near the end of their lifetimes, they return much of this matter to interstellar space, either puffing it out gently like a smoke ring or, if they are larger than about 10 times the sun, in a spectacular explosion known as a *supernova*. About 95 percent of the stars in the Milky Way will end their lives in the gentler manner, shedding mass in a solar wind that can go on for about a million years.[2] Many different compounds have been found in the star's surrounding halo of material, known as a *nebula*. Although the star itself produces atoms, the gentler heat in its environment can create larger molecules from the expelled matter. More than 60 different molecules have been identified by their distinctive spectral wavelengths in stellar nebulae by the Infrared Astronomical Satellite (IRAS). These include linear and circular chains of carbons (some of them quite complex) such as CH_3CN and HC_5N, as well as abundant levels of the already mentioned molecules that are the building blocks of life. Some of these compounds accrete into solid-state particles of dust, including amorphous silicates and carbons. Materials like this have been found in asteroids and comets, and were likely delivered to the early Earth during the formation of our solar system.

Among the most primitive meteorites that have fallen to Earth are a group known as *carbonaceous chondrites* that are believed to retain samples of materials present in the early solar system.[3] A variety has been collected, containing 1.5 to 4 percent carbon compounds as well as much material that is similar to terrestrial rock.[4] A well-studied one, the Murchison meteorite, fell in Australia in 1969. More than 1,000 molecular species have been identified in this meteorite, including amino acids, sugar alcohols, and even some of the bases found in RNA and DNA. We know these molecules are of extraterrestrial origin because they are rich in isotopes that are rare on Earth. Some may be formed during the heat of entry into Earth's atmosphere, but it is clear that organics can survive the fiery fall to the ground.[5]

2. Undersea volcanic vents are a rich source of energy and molecules that could serve as precursors to life. Ammonia and carbon dioxide (CO_2) bubble up with the heated water and can lead to a rich yield of organic molecules.[6] Hydrogen gas exits from these vents and today serves all the energy needs of colonies of bacteria found there. These are at the base of a food chain that includes tube worms, clams, shrimp, and blind fish, in a habitat that is essentially independent of sunlight. This fact vastly increases the range of habitats where life might be possible.

These "black smokers" emit hot water (up to 405°C [761°F]) that tends to be highly acidic (pH 2–3), but a gentler kind of vent system was discovered in the year 2000 that may be more suitable for the origin of life.[7] These exist in shallower water, emit fluid at less extreme temperatures (40 to 90°C [104 to 194°F]) and are alkaline (pH 9–11) rather than acidic. The ones at Lost City in the mid-Atlantic have gradually been building up chimneys of precipitates for at least 100,000 years. These towers are riddled with tiny pores and lined with mineral clusters containing iron, nickel, and sulfur exuded from the hot mantle rock beneath the vents.[8] Research shows that metal containing enzymes are ideal for forming a variety of organic molecules used by life. These gentler vents are host to diverse colonies of anaerobic bacteria that use methane (CH_4) bubbling up from below as their energy source.

3. In the 1950s, Stanley Miller and Harold Urey zapped electric sparks through simple gases they believed would be present on the early Earth— hydrogen (H_2), methane (CH_4), and ammonia (NH_3)—and they obtained a variety of amino acids, the building blocks of proteins, and other organic molecules commonly used by cells. The sparks were meant to simulate lightning in the early atmosphere and the compounds precipitated into sterile water below were meant to represent the ancient ocean. This work was later rejected as it became apparent that the early atmosphere of a planet is more likely to have high concentrations of nitrogen (N_2) and carbon dioxide, which do not yield abundant organic molecules under these conditions.

More recent research has shown that if iron ions are added to the water (Fe^{2+}), there are high yields of amino acids after all. The iron inhibits oxidizing agents that are also produced in these experiments that would otherwise break down the yield of amino acids. It is entirely reasonable to believe the early oceans held abundant levels of iron ions.[9] Such is the fate of scientific ideas that can wax and wane like the stock market.

There are, therefore, at least three sources for the molecules of life: chemical processes in outer space and at hydrothermal vents, and electric discharge through primitive atmospheres. We have not, of course, explained the origin of life, but we can at least find sources for its parts, and I will conclude with a scenario about how these might have come together. But it should be realized that these three sources are not mutually exclusive. The Earth, as well as the whole solar system, condensed from a vast cloud of dust and gas that was drifting through space, and the planet was subject to intense bombardment of material during at least the first 700 million years of its existence. This culminated in a period known as the *late heavy bombardment* about 3.9 billion years ago (bya), an onslaught of meteorites that went on for some 20 million to 200 million years.[10] The pockmarked face of the moon still bears witness to this assault. Recent evidence shows that the Earth may have cooled enough to form a crust as early as 4.3 bya, that water was present on the surface, and that plate tectonics may even have begun then. The evidence comes from zircons in Western Australia dating to between 4 and 4.27 bya. These are rock inclusions that indicate they formed at relatively low temperatures (600 to 780°C [1,112 to 1,436°F]) and under water.[11]

Craters were probably widespread on the early Earth, as they still are on the moon. Many were probably filled with water, and the thinner crust, punched by the impact of meteors, could easily form hydrothermal vents at its base as it cooled. These depressions could also go through cycles of collecting rainwater and partial drying at their edges. All three systems described—shallow hydrothermal vents, material from outer space, and rainwater from electrical storms, each carrying its own load of organic material—could therefore come together in these craters. Millions of experiments may have been going on around the early planet, each with its own combination of precursor molecules, temperature, and pH. Examination of a variety of carbonaceous meteors that have fallen to Earth show that organics survive the fiery passage through the atmosphere,[12] and that conditions around hydrothermal vents can leach them into the water. Also, the heavy bombardment of the Earth by meteors in its first billion years was likely not the globally sterilizing, ocean-evaporating experience once thought. Microfossils of cells go back at least 3.4 bya.[13] At that time, the moon and the Earth were still experiencing about 15 times the impact rate they have today, but this clearly did not exterminate the microbial life then present on the planet.[14]

Studies show that impact craters have high numbers of pores and high levels of permeability in the underlying rock. Clays are also common at these sites, along with iron, iron sulfides, and nickel. In many cases, the metals were brought in by the meteor itself. Compression of the rock by impact often begins a process called *serpentinization*, which produces heat for a hydrothermal vent and also releases hydrogen gas, methane, and hydrogen sulfide from the rock.[15]

The Molecules of Life

I have suggested three possible sources for the molecules of life, but we should look in more detail at how these precursors might be made and what role they play in the life of cells. Any description of the origin of life must be able to account for the source of four different groups of molecules, and an understanding of their role in cells will help explain how they might have come together.

PROTEINS   Much of the structure of cells is made of protein, and almost all its chemical reactions are mediated by protein enzymes. All proteins are made of 20 amino acids, much as all English words are composed of just 26 letters. The particular sequence defines the use, and proteins are specialists for folding into many different three-dimensional shapes according to their amino acid sequence. The side groups of amino acids can bend in two possible directions, called *enantiomers*, that are mirror images of each other, much as a left and a right hand are opposites of each other. A right hand can fit with and shake only another right hand. Life on Earth uses only one of the two possible enantiomers of amino acids, known as the *L form*.

Over 100 different amino acids have been found in the Murchison meteorite, eight of which are in proteins on the Earth. It is interesting that some of these extraterrestrial amino acids show a small but significant enrichment of the L isomer.[16]

A lab experiment set up a high-vacuum chamber with ices of simple molecules found in interstellar clouds, such as water, ammonia, and carbon dioxide, and irradiated them with ultraviolet light, which is abundant in the vicinity of stars. They formed 16 amino acids, six of which are found in Earth proteins. Some were the same amino acids discovered in carbonaceous meteorites.[17]

Spark experiments with gases of the early Earth, as already described, also produce abundant levels of amino acids, provided that iron ions are supplied to the water where the byproducts accumulate.

LIPIDS   All cells are covered with a membrane that is made mainly of lipids. Fats are also lipids that store energy. Lipids have long chains of carbon atoms with hydrogens attached (a "saturated fat" has the maximum possible number of hydrogens attached to the carbon tail). Like fats, most of a lipid repels water (is hydrophobic), but in cell membranes the carbon chain has a polar group at the top that interacts with water. These molecules thus have a double nature (called *amphipathic*), repelling water with their long tails but interacting with water at the polar head (polar groups have some electrical charge to them). Pour a bunch of these molecules into water and they can spontaneously form membrane structures like those surrounding all cells.

Soap is made of long-chained amphipathic molecules, and soap bubbles give useful insights into the dynamics of cells. You can verify this in the bathtub, especially if you take a bubble bath. Two soap bubbles can touch, merge their membranes, and form a larger bubble. A growing bubble can split in two and form two smaller bubbles. Both these dynamics are regularly found in cell membranes. As a cell grows, it adds lipids to its membrane until it splits in two, forming two smaller daughter cells. Cells can meet, merge their membranes, and thus pool their contents. This happens every time a sperm fertilizes an egg and is the way many viruses gain entry to their host cells. (Some strong soaps are so much like cell lipids that they are actually dangerous to handle; they can insert into cell membranes and disrupt them.)

Simple amphipathic molecules have been found in carbonaceous meteorites and are also produced in the lab in a variety of conditions that simulate interstellar ice and hydrothermal vents.[18] The Murchison meteorite contained lipidlike molecules that could form small "bubbles" called *vesicles* by assembling themselves in water.[19] These vesicles could maintain a gradient of molecules inside themselves that was different from their surroundings. In the lab, long chains of lipidlike molecules form from simple gases like carbon monoxide, hydrogen, and carbon dioxide, all of which are abundant in interstellar clouds.[20] Similar results are found in conditions like those surrounding hydrothermal vents. These volcanic vents often release metallic iron, which can act as a catalyst to volcanic gases such as hydrogen and carbon monoxide to form long hydrocarbon chains, up to 30 carbons long, by the stepwise addition of single carbon atoms.[21]

Modern cell membranes have complex channels that allow selected traffic of molecules into and out of cell interiors. Nothing this sophisticated should be expected at the dawn of life. But researchers have found that lipids with tails of 12 to 14 carbons long are semiporous, allowing selected traffic into and out of the bubblelike vesicles they form.[22]

DNA AND RNA   DNA is the master blueprint of cells, carrying all the genes a cell will use. RNA is a copy of one or a few of those genes that a cell is using at a particular time. It is the template upon which proteins are assembled. Modern cells follow the pattern of DNA → RNA → protein; that is, a DNA sequence produces an RNA sequence, which leads to a protein (a sequence of amino acids).

DNA is a double-stranded molecule that uses a simple four-part code: A–T and G–C. That is, A (adenine) always pairs with T (thymine), and G (guanine) always pairs with C (cytosine). Thus, a sequence on one strand determines the sequence on its complementary strand. If one side of the DNA reads ATGCT, the other side reads TACGA. This code is made by four nitrogen-containing bases that point to the interior of the double-stranded helix of DNA, like the rungs inside a ladder. They are attached to a backbone of sugar (ribose) and phosphate that runs along the outside of each side of the helix.

RNA is a single-stranded molecule with a structure similar to one side of the DNA helix, and it also uses a four-part code: A, U (uracil), G, and C. It can pair with one side of the DNA just as DNA's own complementary strand can, except RNA uses A–U pairs whereas DNA uses A–T pairs (uracil in RNA has a very similar structure to thymine in DNA). When a cell wants to activate a gene, it opens up that area of DNA, makes a copy onto RNA, and then uses that RNA to produce a protein, much as an architect might photocopy one page from the master blueprint and send it out to the construction site to be built.

Every triplet of RNA specifies a particular amino acid. Thus, CAG on RNA specifies one of the 20 amino acids, and AGC specifies a different amino acid. Three of the triplets mean "stop, end of message," like the period at the end of a sentence.

Because RNA is single-stranded, it can also fold over on itself and bind in selected places if the sequence in one area is complementary to a sequence in another area. In this way it can form complex three-dimensional shapes that have functions besides just carrying information. A group of RNAs known as *ribozymes* can act as catalysts for other chemical reactions. One ribozyme, for example, can catalyze the copying of another strand of RNA to produce a complementary copy. The main enzyme that hooks separate amino acids together into proteins in modern cells is also an RNA found inside a complex subunit known as a *ribosome*.[23]

These diverse talents of RNA have led many researchers to suggest that early life may have been an "RNA world," where RNA both carried genetic information and did enzymatic activities, as many modern protein enzymes do. Only later in evolution did DNA come along as a specialist for storing information and protein as a specialist for three-dimensional shapes. In the modern world, we see RNA as an intermediary between these two realms, but in a simpler, less competitive world, it may have performed both functions.

RNA is not easily produced from simple precursors, although we now have some strong indications of the process. All four bases of RNA have been synthesized from simple molecules. Adenine (A) can be made from hydrogen cyanide (HCN) and ammonia (NH_3),[24] both of which are found in the interstellar medium.[25] The Murchison meteorite carried bases of RNA.[26] Ribose, the sugar in the backbone of RNA and DNA, is easily produced from formaldehyde (CH_2O), which is found abundantly in both comets and interstellar space.[27] The problem was seeing how the sugar and base could come together, but a pathway has now been found using inorganic phosphate as a buffer early in the reaction. Phosphate is present in volcanic gases,[28] and all the starting materials for forming RNA nucleotides could plausibly have been present under early Earth conditions.[29]

SUGARS   Used in cells for energy, sugars are also part of the backbone of RNA and DNA and can form other stable structures as well. For example, cellulose, the main component of wood, is a polymer of the sugar glucose. All sugars are some multiple of the atoms CH_2O. Glucose, for example, has six multiples: $C_6H_{12}O_6$, and ribose has five multiples: $C_5H_{10}O_5$. These and other sugars are readily made from formaldehyde (CH_2O), which is found abundantly in comets, meteorites, and the interstellar gas clouds where solar systems are born. Sugars and sugar alcohols ranging from three to six carbons have been found in carbonaceous meteorites in amounts comparable to the amounts of amino acids found in those bodies.[30] Sugars are also produced under conditions around shallow hydrothermal vents where ultraviolet light can make formaldehyde from carbon dioxide and water, which in turn can form sugars under alkaline conditions.[31]

All the precursors for life could, therefore, have been readily available on the early Earth. There is nothing unique about the chemistry described here, and the precursors all use atoms that were and still are abundant in the interstellar medium. How they might have combined into the earliest life forms is harder to conceive, and I will present one of the more likely scenarios. No one can yet definitively answer this question, and to some it may seem like a miraculous event, only possible by the hand of God. But if it were a divine event, discontinuous with the rest of nature, we should not expect so many hints, so many inroads, to a solution to the question. It would stand in stark contrast to the rest of nature. Instead, we see continually rising ground that indicates a peak ahead that we have not yet scaled. We have partial answers that are becoming more complete year by year, and we may one day be able to piece the whole story together.

Pathways to Life

I have suggested impact craters as ideal sites for experiments leading to life on the early planet. As these cooled, they could accumulate water that was delivered to the Earth by comets or rainfall. The Earth was subject to the same bombardment as the moon, and therefore its face too would have been pockmarked with these depressions (which later plate tectonics would erase on our world). In some, shallow hydrothermal vents could form, either by rupture of the early thin crust to the magma below, or by serpentinization of impacted rock. These craters therefore could be repositories of organics from the three sources already described: meteors, hydrothermal vents, and rainfall. Repeated cycles of drying and wetting could take place at their edges, which would also help concentrate their contents. Some would have existed as isolated pools or lakes, whereas others could have been gathered up by an expanding ocean.

Hydrothermal vents have highly porous precipitates of calcium carbonate that build up chimneylike structures on top as the vent ages. The sidewalls are riddled with pores the diameters of which vary in size from millimeters to micrometers. A series of lab experiments using minerals from hydrothermal vents found that narrow pores of this size can accumulate molecules that flow past them. A test with single nucleotides found that they could concentrate 100 million-fold in a space about the diameter of a cell.[32] Longer strands accumulated at higher concentration than single nucleotides, and at the bottom of the pore, the molecules "diffuse freely and would find chemical reaction partners comparable to the situation within prokaryotic cells."[33] Linking pores furthered this process, so you might have an assembly line of different pores in the sidewalls of chimneys with different concentrations of molecules and at different temperatures as the pores grew away from the vent's heat source. The researchers believe this process would work equally well to accumulate amino acids or lipids flowing by. The concentration process takes minutes to hours, whereas the Lost City vents, from which these minerals were taken, have been at their ocean locations for at least 30,000 years.[34]

Moreover, these pores are lined with a film of sulfur and metal clusters that are regularly exuded from the underlying heated rock through the chimney vents. In modern cells, metallic or metal-sulfide catalysts are involved in many of the key steps of metabolism, including the synthesis of amino acids, proteins, pyruvate, and the reduction of nitrogen to ammonia. The clusters found on some iron- and sulfur-bearing minerals have a geometry similar to that found in the active sites of modern biological enzymes.[35] In the early days of life's origin, before cell competition became intense, these naked clusters may have been sufficient to produce many biological molecules from the simple gases bubbling up from hydrothermal vents. Iron sulfur (FeS) and nickel sulfur (NiS) by themselves can catalyze biochemical reactions to produce organic molecules, using hydrogen gas as an electron donor and carbon dioxide as an electron acceptor.

The pores of hydrothermal vents, then, may have been the first site of the "RNA world" postulated by many theorists. One RNA strand can serve as a template for the assembly of a complementary RNA strand, which is then a template for a copy of the sequence on the original strand, and so on. A ribozyme, an enzyme made of RNA, can catalyze the joining of single RNA nucleotides into chains. The ribozyme folds over into a three-dimensional shape and clutches two metal ions to do its work.[36] In one experiment, a ribozyme, E1, catalyzed the connection of two lengths of RNA to make a second ribozyme, E2. E2, in turn, connected two lengths to make more E1, and the system grew with exponential amplification as long as more RNA subunits were available.[37]

The enzymatic activities of RNA can be enhanced by their association with amino acids. Amino acids make up the three-dimensional shape of proteins, and amino acids associated with RNA could enhance its specificity for other three-dimensional shapes, increasing its binding ability and enzymatic effectiveness. In modern eukaryotic cells, RNA-protein associations have important catalytic functions in fundamental components (organelles) of the cell. These include joining amino acids together to build proteins (in ribosomes), matching the proper amino acid to its tRNA (by tRNA synthetase), editing out the unused sections of genes (by a spliceosome), and extending the ends of chromosomes (by telomerase). In each case, shapes created by RNA folding are enhanced by their association with the amino acids of proteins. Some RNA chains have a natural affinity for particular amino acids. A study found eight amino acids bound by simple RNA chains that recognized distinct features on the individual amino acid. These RNA chains had distinct triplets of RNA near the binding sites, and the authors suggest this might be the beginning of the triplet genetic code that defines all amino acids in modern cells.[38]

There are also organizing effects that take place between RNA and lipids. Nucleotides of RNA were mixed with lipids in water and put through repeated cycles of drying and wetting. This process, along with heat (up to 90°C [194°F]) in water, was sufficient to create chains of RNA, mostly in the range of 25 to 75 nucleotides long. At the end of the reaction, RNA chains were found inside lipid membranes, much as cell membranes hold genetic material in modern cells.[39] Repeated cycles of wetting and drying could be expected at the borders of craters on the primitive Earth. Lipids are also known to form at hydrothermal vents, where they assemble into vesicles that grow and split into two more vesicles in dynamics similar to those of cell membranes splitting into two.[40] It is harder to see how cycles of wetting and drying would affect the pores inside vent chimneys, but this would have been possible in shallow vent pools that were subject to periodic drying and rainfall. As mentioned earlier, millions of different experiments could have been occurring on the early Earth in pools of different sizes as well as at vents under the burgeoning ocean.

Origins of Metabolism

The last major factor necessary for life is an energy source. All life we know of has a complexity that requires an input of energy to build and maintain its structures. The early Earth had abundant energy sources, from ultraviolet light to lightning to volcanic activity. Hydrogen gas, the most abundant molecule in the universe, was part of the swaddling clothes of the young planet, before its lightness made the gas drift off into outer space. Hydrogen gas is also released as rock is oxidized by water in the process called *serpentinization*. In addition, carbon dioxide was abundant on the young planet, acting as a greenhouse gas to keep it warm. There was little free oxygen as yet, for it is highly reactive and would have been bound up in compounds until set free by photosynthesis later.

The ancient oceans are estimated to have held a thousand times more carbon dioxide than at present.[41] Carbon dioxide dissolved in water produces carbonic acid, giving water an excess of protons (H+). At the same time, serpentinization at hydrothermal vents produces alkaline effluents (a deficit of protons) along with hydrogen gas. Thus, there was a pH gradient across the walls of the vent chimneys, an excess of protons (H+) outside, and a deficit inside. It turns out that all modern cells use H+ gradients for energy. The flow of protons through a complex enzyme assembly known as *ATP synthase* creates adenosine triphosphate (ATP), the most common energy currency used by all cells. Adenosine triphosphate is an RNA nucleotide (figure 10.1), the A of the four-part RNA code. Like all nucleotides, it ends in a tail of three phosphate groups. These phosphate groups have negative charges on them, all cramped together in a narrow area, so it is a system under tension. When the bond to the last phosphate group is broken, some of that tension, like a coiled spring, is released. The released energy is used to

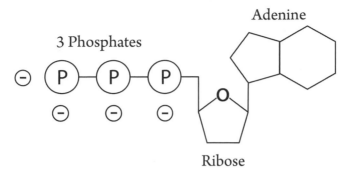

Figure 10.1 ATP (adenosine triphosphate).
ATP is a typical nucleotide with three components: a coding base (in this case adenine, or A), a sugar (ribose), and three phosphate groups (a triphosphate). The negative charges on the phosphate groups repel each other and are a source of tension. When one or more phosphate groups break off, energy is released. ATP can be used either for energy or as a coding base in RNA.

power all kinds of chemical reactions around cells, as well to energize a variety of mechanical movements. The result is ADP (adenosine diphosphate) and a free phosphate group (called *inorganic phosphate*, or *Pi*).

The spring has to be recoiled to provide more energy by regenerating ATP. All cells use the energy of H+ gradients to put ADP + Pi back together again as ATP. We use approximately our body weight of ATP in a day, yet we store only about 50 grams (1.76 ounces) of the molecule. This means that each ATP molecule gets recycled about 10,000 times a day in our bodies.[42]

ATP synthase, which rebuilds ATP, is a complex, membrane-spanning enzyme, and it is unreasonable to assume anything that complex existed at the dawn of life. There are simpler versions of this mechanism, however. Two copies of a membrane-spanning protein (a dimer) are able to use an H+ gradient to put two phosphate groups together (called *PPi*), creating an energetic molecule with characteristics similar to ATP. Sequences of this protein are highly conserved in bacteria and plants, so it may be very ancient.[43] In fact, an H+ gradient by itself across a metal sulfide membrane, as might be found lining a hydrothermal vent, can create a diphosphate from individual phosphate groups.[44]

Splitting phosphate groups for energy may have appeared very early in the history of cell metabolism. In an RNA world, the tails of nucleotides such as ATP could have easily assumed this function also. Indeed, many of the key transfers of energy in all cell metabolism today are mediated by nucleotides, or modified versions of them, such as ATP, GTP, NAD, NADP, FAD, and CoA.[45] Coenzyme A (CoA) is an extended nucleotide with a sulfur group on its business end. This forms a high-energy bond with acetate, a central molecule in the Krebs or tricarboxylic acid (TCA) cycle. Some cells, such as those of the green sulfur bacteria, can run this cycle in reverse, producing subunits for all the molecules of life: lipids, sugars, amino acids, and the nitrogenous bases of nucleotides.[46] ATP and GTP, the main energy currency in all cells, are built the same as the single nucleotides (A and G) used for coding purposes in RNA. NAD and NADP, key shuttles for high-energy electrons in metabolism, each have two nucleotides stacked on top of each other. The fact that all these forms of nucleotides play key roles in energy transfers in modern cells may be a remnant of their pivotal roles in the earliest metabolism of cells. In a simpler RNA world, nucleotides had the potential to not only code for information, but also to act as enzymes and facilitate energy exchanges.

Escape from the Vents

Putting this information together, we can sketch out the following rough scenario for the origin of life. It is one of many possible sequences, but, I believe, one of the most likely, with the proviso that future research will make these issues clearer. Life may have begun at mild hydrothermal vents where an alkaline effluent met colder acidic water. These vents could have been in shallow seawater or at the base of depressions left by meteor impacts. Monomers for life molecules were already present, either brought in by meteors, formed by lightning in the atmosphere, by the chemistry at hydrothermal vents, or some combination of all three. Metals and sulfur exuded from the vents lined the interiors of mineral chimneys that formed above the vents. Within these chimneys, tiny pores were able to accumulate and concentrate organic molecules, forming, in effect, tiny test tubes with the width of cells. The walls of some of these tubes could be lined with metal-sulfur clusters that are commonly exuded at hydrothermal vents. These have the ability to act enzymatically to form early polymers. Life begins as an RNA world, with short sequences of RNA that are able to code for daughter strands and also to fold into RNA enzymes (ribozymes) able to catalyze the joining of single nucleotides into chains. The association with amino acids enhances the three-dimensional specificity of these RNA enzymes, and particular sequences on RNA associate preferentially with particular amino acids—the beginnings of a genetic code.

Vent pores form an interconnected series of chambers that lead away from a central underground heat source. Although heat from the vent might have been sufficient to provide activation energy for early chemical reactions, the farther a mixture traveled from the center, the more it would need high-energy intermediates to spur chemical reactions. These were provided by phosphate groups from diphosphate (PPi) at first, and then the triphosphate tails of RNA nucleotides. Phosphate polymers were regenerated by the H+ gradient that naturally exists between the alkaline interiors of the chimneys and the acidic surrounding waters. The first enzymes to enhance the formation of bonds between phosphates would also be embedded in the metal sulfur films lining the interior of the pores.

The next stage of this process would be a further enhancement of the RNA-amino acid association by enclosure within a lipid membrane. Although the diameter of a pore itself was enough to concentrate the chemical reactions, they would be more mobile when enclosed within a membrane. Lipids are formed abiotically at hydrothermal vents and are able to organize and enclose RNA, as we have already seen. As the lipids formed enclosures around the first protocells, they may have taken some of the H+ conducting enzymes from the pore walls with them. These would continue to join phosphate groups for the cell as an energy source using H+ gradients.

Later evolution of these cells would entail the development of DNA from RNA (there are only small chemical differences in the sugar and coding bases between them). An enzyme called *reverse transcriptase* can use an RNA template to build DNA. Elsewhere in the protocell, the RNA ribozyme that joined amino acids together into short proteins would be further enhanced by its association with amino acids until a full-fledged ribosome evolved that was efficient in building proteins. In modern cells, ribosomes are still made of combinations of RNA and protein, and their catalytic function of joining amino acids together is performed by a short internal RNA only 29 bases long.[47] Thus we arrive at the division of labor found in cells today, where genetic information is stored in DNA, three-dimensional structures are built mainly of protein, and RNA acts as an intermediary between these two realms. Only after a cell had acquired all these functions would it be able to leave the nursery of the hydrothermal pores and live in the surroundings waters. Such a cell is imagined to be the last universal common ancestor (LUCA), as the mother of all cells that came afterward.[48] The universality of the genetic code, the structure of ATP synthase, and the core of the translation system that produces proteins indicate that all cells go back to a common source.

Clearly, there are gaps in this scenario, places where complex steps have been glossed over with a phrase and some of the transitions are still mysterious. Much further research is needed to flesh out this narrative. But there was world enough and time for these many evolutionary steps to take place. The early Earth was likely dotted with impact craters and hydrothermal vents all over its surface. It is possible that it had cooled sufficiently to allow these experiments in biogenesis by 4.3 bya.[49] The first unambiguous microfossils of cells date back to about 3.5 bya, and these were already complex bacterial colonies.[50] Somewhere in the intervening 800 million years, life gained a foothold on the planet. A modern bacterial cell can go through its entire life cycle in 20 minutes. Life may have begun very simply and crudely, but there were huge intervals of time to improve its act through repeated cycles that had small changes in them. One of the beauties of evolutionary logic is that it does not require things to be perfect. You only have to do better than the competition. As long as things work at a bare minimal level, the competitive process can gradually optimize the mechanisms.

Nor was there likely steady improvement in a single lineage. Before LUCA left the vents, perhaps even before it acquired a proper cell membrane, there was probably a lot of promiscuity among gene segments.[51] A study of more than 500,000 bacterial genes found that about 80 percent of them have been involved in horizontal gene transfer—that is, insertion of segments from other species.[52] Modern bacterial cells have a variety of mechanisms by which they can take up foreign DNA and insert it into their own genomes. Viruses may also have been instrumental in these exchanges. There are both RNA and DNA viruses, and some have the ability to insert into the host's genetic material. In the many pores believed to have existed in the ancient hydrothermal vents, it would have been easy to exchange RNA segments and insert them into growing chains. In one pore, perhaps, there were improvements in metabolism, in another the beginnings of lipid associations, and so forth. LUCA emerged from a confederation of segments that were efficient at energy extraction and fidelity of copying and could live independently of the nursery of pores. It probably left many also-rans behind.

LUCA was likely a chemoautotroph, still dependent on energy-rich molecules from the vents, such as those with hydrogen gas. It could make use of high-energy electrons, combine them with carbon dioxide, and build up all the organic molecules it needed by processes such as the reverse TCA cycle.[53] Only much later did true independence arrive with the evolution of more complex systems like photosynthesis. But cells never got over their appetite for H+ ions (protons). In photosynthesis their source has been transferred to water. Using pigments and sunlight, electrons from water (H_2O) are boosted to high energy levels. Water splits, bubbling off oxygen, and the remaining H+ is shunted through an ATP synthase, providing ATP for the cell. Meanwhile, the high-energy electrons are given to a shuttle, NADP+, one of the double-stacked nucleotides described earlier. They are carried to the Calvin cycle, which takes carbon dioxide out of the air, and builds up organic molecules for the cell as part of photosynthesis. Thus an ancient story is told again, in a much more sophisticated way.

The scenario just described is one version of the story that can be told. Some theorists propose that metabolism came before coding, others think membranes must have come first. Different scenarios place the origins in warm pools, deep undersea vents, outer space, or other planets.[54] Much further research will be needed to resolve these questions. We have scenes from a drama—in some cases whole acts, in other cases only snatches of dialogue, but we are on the way to fitting the whole play together.

But it is clear that life began under conditions that would hardly be called hospitable. The average temperature of the Earth was hot, up to 500°C (932°F). The onslaught of meteors was at least 15 times what we now experience, and life may have even survived the late heavy bombardment that lasted until 3.9 bya, when the impact rate was much higher. The earliest unambiguous evidence of life goes back to 3.49 bya, and a little later, at 3.4 bya, microfossil colonies that were likely photosynthetic have been found.[55] This shows that microbial life was already quite diverse, and photosynthesis requires the interaction of several complex systems that would have taken a long time to evolve. It is reasonable to assume that LUCA was much older than this. A study of protein sequences from 72 species of bacteria, using the molecular clock method, dates the earliest archaebacteria to 4.21 bya, with photosynthesis evolving much later.[56] There is controversy about how accurate this comparative method is. But recent evidence indicates that the very early Earth could have supported conditions leading to life.

The Earth formed about 4.6 bya, condensing from a cloud of dust and gas that also formed the sun and other planets. In the earliest days, the Earth apparently had an evil twin, another planetesimal forming near its orbital distance from the sun. Both bodies could not survive; one of their gravities would prevail. The other body crashed into the Earth about 4.55 bya, creating massive debris that would condense into our moon.[57] Yet recent evidence from zircons in Western Australia indicates the presence of a solid crust and liquid water on our planet by at least 4.3 bya.[58] From that time on, hydrothermal vents, such as those described above, would have been possible. Calculations show that even under the heaviest bombardment of remaining meteors, less than 10 percent of the Earth's surface would have experienced temperatures above 500°C (932°F).[59]

Studies of bacterial sequences in the different domains of life indicate that hyperthermophiles, cells that prefer to live at high temperatures, lie at the root of the tree of life.[60] An archaebacterium that lives near a deep hydrothermal vent in the Pacific doubles its numbers at 121°C (249.8°F) and ceases growth below 85°C (185°F).[61]

Not only did life start under harsh conditions, but it has crept into niches where its existence would have been considered impossible until recently. Life persists despite extremes of temperature, acidity, pressure, radiation, dryness, and salinity.[62] In a mine shaft in South Africa, at 2.8 kilometers (km [1.75 miles]) down, in alkaline, saline ground water, researchers found a bacterial species that lives off sulfate and hydrogen and does not rely on any products produced by the sun.[63] Some bacterial species may exist as far as 5.3 km (3.29 miles) below the surface.[64]

Antarctica has some of the coldest, driest areas on the Earth, yet various bacteria, lichens, and algae live there. When NASA scientists drilled below 180 meters (590 feet) of Antarctic ice, they found a shrimplike animal about 7.6 centimeters (3 inches) long in an area that does not see light and is more than 19.3 km (more than 12 miles) from open water. Elsewhere in Antarctica, water that averages only −1.8 to 1°C (28.76 to 33.8°F) is teeming with life, including clams and bacterial colonies 0.8 km (about one-half mile) below the ocean surface.[65]

The ease with which biological precursors are formed and the early appearance of life on our planet under harsh conditions suggest that life may be an accident waiting to happen, given the right circumstances. Life may be a natural state of matter, a special case of the self-organizing structures that emerge at energy gradients. It is a kind of dissipative structure that has gained heredity as a way to persist.

There are bulls and bears on the prospects for life, as there are for the stock market. Palaeobiologist Simon Morris, whose views on the evolution of intelligence are similar to the ones presented in this book, nevertheless thinks conditions for a habitable planet must be rare, and he has titled one of his books *Life's Solution: Inevitable Humans in a Lonely Universe*.[66] However, biochemist Christian de Duve, a Nobel laureate, thinks the universe must be "awash with life" and has titled one of his books *Vital Dust: Life as a Cosmic Imperative*.[67] In the next chapter we will examine the prospects for worlds like our own within the habitable zone around stars, where life might begin its evolutionary journey.

11

The Prospects for Habitable Worlds

The idea that there may be planets around stars other than our sun was once the stuff of science fiction, but research in the last two decades indicates that such planets may be a common occurrence. Data from the most recent space telescopes indicate that even Earth-size planets in the habitable zone around their star may be numerous in the universe. Both theoretical studies and the latest data from space telescopes seem to confirm the Copernican principle that there is nothing special about this quadrant of the galaxy and conditions that led to a solar system here may be present all over the cosmos.

Most astronomers now believe that planet formation is a regular part of star formation.[1] Stars and planets develop together from massive clouds of dust and gas that float through interstellar space. These clouds are composed mainly of the two primordial elements formed at the Big Bang, hydrogen and helium with a small enrichment of heavier elements produced by stars that have burned and expired since the Big Bang. These clouds are floating through the galaxies of which they are part, and they often have at least some slight rotation to them. If an area begins to compress through gravitational attraction, that rotation will speed up, much as a turning skater spins faster as she pulls in her arms due to the conservation of angular momentum. As the cloud condenses, a disk will form around the equator of its central region because this is where centrifugal (outward) force best counters the increasing inward attraction of gravity (figure 11.1). Most of the mass will be concentrated in the center, where a star will be born, and planets condense from the residual matter left over in the rotating disk. This model explains many

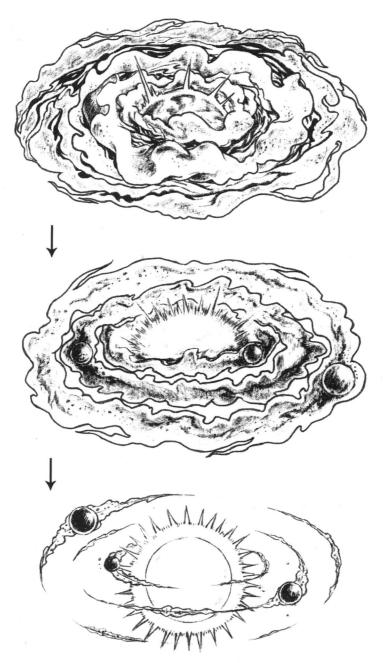

Figure 11.1 A solar system condensing from an interstellar cloud of dust and gas.
A disk forms around the central mass that will ignite to become a star, and a retinue
of planets condenses from residual material in the disk.

characteristics that we find in our own solar system. All the planets orbit in nearly the same plane at our sun's equator, and they move in nearly circular motion in the same direction as the sun's own rotation, indicating a common birth from a larger system.

This pattern is repeated in the moons of the solar system, most of which orbit their planets in the same direction as the planets themselves turn.[2] The largest planets—Saturn, Jupiter, Uranus, and Neptune—also have rings orbiting their equators with particles moving in nearly circular motion.[3] The moons of Jupiter have been called a miniature solar system. They move in roughly circular paths near the plane of the planet's equator. The innermost moons are rocky, like the inner planets of our solar system, and the outmost ones are less dense, with a large fraction of ice and water as is found in the outer giants of our solar system.[4]

This pattern of condensation appears to apply also on a larger scale, in the shape of many galaxies that have a luminous core of densely placed stars surrounded by a disk of stars that are moving in roughly circular orbits, in the same direction and in nearly the same plane. The disk of our Milky Way is about 100,000 light-years across but only 1,000 light-years thick,[5] and it takes about 230 million years to complete one rotation (the last time we were in this galactic quadrant, dinosaurs roamed the Earth).

This pattern of disk formation is so ubiquitous that it has been found around brown dwarfs, failed stars just below the lowest limit for igniting nuclear fusion;[6] around neutron stars, which are the remnants of stars that exploded as supernovas;[7] and at the outer edges of black holes, where matter swirls in a disk before being swallowed down the throat of an invisible monster.

A survey of 84 stars near the sun found that most stars younger than 300 million years old have disks of dust and gas, but that 90 percent of those 400 million years old do not.[8] Where do these disks go? The most likely explanation is that particles in the broad band of a disk radiate enough infrared light to be detectable, but when they are condensed into a small number of planets, they do not give off enough light to be seen. Until recently, our instruments were too crude to detect almost all planets smaller than the size of Jupiter outside our solar system.

Although the role of disks in planet formation was debated for many years, their continuity was confirmed by a study of a nearby sunlike star. Epsilon Eridani has a mass of 83 percent of our sun and is less than a billion years old. It is surrounded by a disk of dust and gas stretching out to about 65 times the distance of the Earth to the sun. A planet has been confirmed within that disk with a mass about 1.5 times that of Jupiter. Both the disk and the orbit of the planet lie in a plane tilted at 30 degrees to the viewer on the Earth.[9] Several other nascent solar systems are also now known to contain both planets and debris disks, although our instruments are still only at the edge of being able to detect both these features.[10]

Our solar system formed about 4.6 billion years ago (bya), in a universe that was already some 9 billion years old and had gone through several generations of star formation. The primordial cloud from which our solar system formed was still 98 percent hydrogen, helium, and a trace of lithium—elements left over from the Big Bang—with about a 1.6 percent enrichment of heavier elements strewn into space from expired stars that had lived earlier. This small admixture from stars was mainly metals, rocky material (primarily silicates), and compounds formed with hydrogen, such as water, methane, and ammonia.[11]

As our solar nebula began to condense, it heated up through the conservation of energy: gravitational energy was converted into the kinetic energy of moving particles, and as these began to collide with one another, they heated up. When the center, where the protosun was forming, reached a temperature of about 1,300°C (2,372°F), a temperature gradient stretched out to the edges of the gathering disk. Near the sun it was too hot for anything but metals to condense into particles. Farther out, where Venus, Earth, and Mars were born, rocky materials could also condense, and farther still, past what is known as the *snowline*, ices of water, methane, and ammonia could form.[12]

As the solar disk flattened and condensed, these particles crashed into one another, forming ever-larger chunks over a period of millions of years. Large bodies fragmented smaller bodies and then gathered them up through gravitational attraction, so the rich got richer and the poor got poorer, forming, at last, two kinds of planets. Rocky planets, from Mercury to Mars, formed nearer the sun, where metals and rocky silicates first condensed. Numerous ices past the "snowline" built the gas giants, from Jupiter outward, which—because of their large size—were able to retain some of the lightest elements, hydrogen and helium, in their atmospheres. But the low temperatures at these outer planets also allowed the condensation of the heavier metals and silicates. It is believed that in the depths of Jupiter is a rocky, metallic core.[13]

This pattern of planet formation is considered quite general. Repeated simulations using supercomputers to model the mutual gravitation between many bodies in a nascent star system often show rocky planets rich in water forming near a newborn star. Results vary between one to four terrestrial planets, depending on the distribution of matter in the original protoplanetary disk and the distance of large Jupiter-like planets in the outer regions.[14]

Also, planets may not be in the nearly circular orbits often found in our solar system. Depending on the shape of the original protoplanetary disk, the pattern of collisions between planetesimals, the gravitational influence of planets on each other (resonances), and even the early migration of planets inward toward their star, orbits can be pulled to elliptical rather than circular shapes and may even be above and below the plane through the star's equator.[15]

The Role of Star Size

Stars vary in size from about 8 percent of the sun's mass to 150 solar masses. At the lower limit, they do not have enough material to ignite nuclear fires, and at the upper limit, they burn so furiously that they blow away additional mass that would make them larger.[16] The life and death of stars is related to their size. A star of 10 solar masses burns its fuel 10,000 times faster than our sun and lives only $\frac{1}{1000}$ as long, about 10 million years, while a star of 0.3 solar mass burns at $\frac{1}{100}$ the rate and lives 30 times longer than the sun, about 300 billion years.[17] Stars larger than about two solar masses would not burn long enough to provide the billions of years of stable energy that the evolution of life requires, at least as far as we know it. An F-class star at 1.6 solar masses has a stable output for 3 billion years; a G-class star like our sun, for 10 billion years; and a K-class star at 0.79 solar mass, for 15 billion years.[18] Luckily, the proportion of stars goes up the smaller they are: for every star larger than 10 solar masses, there are 10 stars of between two and 10 times the sun's mass, 50 stars at 0.5 to two times, and 200 stars at 0.08 to 0.5 times.[19]

The habitable zone around a star is defined as the area where water would remain liquid. Too close to a star, water will evaporate and ultimately be lost to a planet through photodissociation in its upper atmosphere. Too far from the star and water freezes. Because water is a universal solvent for life as we know it, it is reasonable to assume it should be necessary in habitable zones. James Kasting defines the habitable zone around our sun as being between 95 and 165 percent of the present distance from the Earth to the sun (0.95 to 1.65 astronomical unit [AU]).[20] Every star will have its potential habitable zone, the Goldilocks region that is not too hot or too cold for water to remain liquid. The larger and hotter the star, the broader and farther out this zone will be. An M star at 0.1 solar mass has a habitable zone only about a tenth as wide as that of our sun, and the ring is much closer to the smaller star. An F star that is about three times as luminous as our sun will have a habitable zone about twice as far out and it will be twice as broad.[21]

The largest stars, at greater than 10 solar masses, play an important role in the history of life. They end their lives in a supernova explosion, blasting most of the material they have synthesized in their cores out into space. Because the Big Bang provided only the lightest elements— hydrogen, helium, and a trace of lithium—all the heavier elements, of which terrestrial planets and life are mainly composed, had to be synthesized inside stars. We are truly star children, and the atoms of our hearts once circulated in the interiors of stars. Stars smaller than about eight solar masses expire in a gentler way,[22] puffing their products into space in giant smoke rings called *nebulae*. A star the size of our sun will return about half its mass into space in this way; a supernova will return almost all its contents to interstellar space.[23]

The most abundant element formed by star burning is oxygen (O_2).[24] Combined with the ubiquitous hydrogen, oxygen ensures that water is plentiful all over the universe. The second most abundant atom is carbon, which is ideally suited for organic chemistry because it can form so many bonds at so many angles. It has become the backbone of life because it can form so many complex molecules with different shapes. It is interesting that life on Earth makes use of the most abundant elements that are chemically active in the cosmos: hydrogen, carbon, oxygen, and nitrogen.[25]

Heavier elements were quickly added to the interstellar medium after the Big Bang. The earliest stars are believed to have formed when the universe was only a few hundred million years old and still retained much of the heat from its fiery birth. Only massive stars would have enough gravity to pull this hot material together. These stars burned through their fuel quickly and returned nearly all their contents to space in the form of elements that might have made planets possible within the next generation of stars. The oldest stars we can now see, at

12 billion years old, are enriched only 0.1 percent by heavier elements, whereas our sun has about 1.6 percent.[26] The Hubble telescope recently spotted a galaxy that is believed to have formed within 600 million years of the Big Bang and that contained about 1 billion stars.[27]

Red Dwarfs

Dwarf stars with masses between 8 and 80 percent of our sun are known as *M-class stars*. Surveys show that most of the stars in the disk of the Milky Way are M-class stars, and they make up 70 percent of our solar neighborhood. They shine with a luminosity of only 0.1 to 6 percent of our sun and can go on burning in a stable way for hundreds of billions of years.[28] A red dwarf at 10 percent solar mass will burn for 5.74 trillion years, and the smallest star, at 8 percent solar mass, will live for 12 trillion years.[29] Our sun, by comparison, will shine as a main sequence star for about 10 billion years, a thousand times shorter than a small M star.

Theoretical studies, as well as observations, show that planetary systems can form around long-lived dwarf stars. Disks are found around even the smallest stars, and they appear to evolve by gravitational collapse in a manner similar to that of more massive stars.[30] A theoretical study predicts that planets about a tenth the mass of the Earth could commonly form around low-mass stars at distances from 5 to 50 percent of the distance from the Earth to the sun.[31] Observational studies show that planets between five and 15 Earth masses may also be common around M dwarfs. An M star with half the solar mass has been discovered with a system that resembles a scaled version of our solar system, containing two large planets (71 percent and 27 percent of Jupiter's mass) at distances proportional to those of Jupiter and Saturn in our solar system.[32] The lifetime of protoplanetary disks around these stars appears to be two or three times longer than around sunlike stars. Their disks are flatter, with dust settling in a way that may enhance planet formation.[33]

Although planetary systems around M stars may be common, there are problems with their potential to harbor life. The habitable zone around a low-radiance star will be closer and narrower than it is around a larger star like the sun. Also, M stars are unusually active in their youth, sending out solar flares and large quantities of high-energy particles. These may be harmful to any genetic system that tries to evolve in its neighborhood. On the other hand, studies show that a thick atmosphere could shield surface life on a close world from all rays except ultraviolet.[34] This is, in fact, a problem here on Earth, and cells have evolved repair mechanisms to fix ultraviolet damage to DNA.[35] Life is unbelievably tough and has found ways to survive in extremes of temperature and radiation. Also, many M dwarf stars settle down after their wild youth to more sedate burning.[36]

A further problem is that bodies in near orbit to larger masses tend to become tidally locked, always presenting one face to their larger companion. The moon does this to the Earth, as does Venus to the sun.[37] It has been estimated that a tidally locked planet around an M star might have its dark side (facing outer space) perpetually frozen, inhibiting the potential for life. But here again a substantial atmosphere could come to the rescue. With sufficient carbon dioxide (CO_2) in the air, winds could distribute the heat, allowing water to remain liquid all over the planet (life might migrate to the dark side to rest).[38] Moreover, tidal locking is not a foregone conclusion. Mercury, the closest planet to the sun, rotates three times for every two of its orbits.[39] A planet with slow rotation is also unlikely to have much of a magnetic shield. Liquid metals must rotate in a planet's core to set up the dynamo that surrounds it with a magnetic field. Without this shield to divert charged particles, more radiation is likely to affect the planet, but here again a thick atmosphere might be sufficient to intercept most of this energy.[40]

Moons and Failed Stars

Planets may not be the only places where there is sufficient energy flow and an environment to potentially harbor life. Giant planets like Jupiter can have systems of moons that resemble small solar systems. Jupiter has four large moons, and the biggest, Ganymede, is larger than the planet Mercury. The gravity of these moons tugs on each other and, together with the more massive attraction of Jupiter, pulls them to somewhat eccentric orbits. This variable force churns the interiors of the moons, creating frictional heat.[41] For the innermost moon, Io, this is sufficient to produce active volcanoes, some of which are erupting almost constantly, as imaged by the Galileo spacecraft that went into Jupiter orbit in 1995.

More intriguing is the next large moon farther out, Europa. It is believed to be covered with ice, surrounding an ocean some 80 kilometers (50 miles) deep, with a rocky core at its center. Tidal heating could be sufficient to create "black smokers" in the depths of that ocean, similar to the volcanic vents where life is believed to have started on the early Earth. On our world, energy-rich molecules spewing from these vents are sufficient to support an entire ecosystem that is essentially independent of light from the sun.

Thus, an energy gradient created by tidal heating potentially could support life on other bodies where starlight is scarce. Giant planets have been found in orbit around M stars. As noted above, such planets are likely to be tidally locked so they always present the same face to their star. Moons around such a planet would be tidally locked to their planet rather than to the central star and therefore would present a variable face to the star, leading to more equitable heating of the moon. Thus moons of a large planet around an M star may be a favorable locale for the emergence of life. Theoretical studies show that moons down to 12 to 23 percent of the Earth's mass could maintain habitable environments if their planets orbit within the habitable zone of their star.[42] Similar arrangements might exist around brown dwarfs, failed stars that are below 8 percent of the sun's mass. Infrared surveys show they are numerous in our galaxy. Giant planets have been detected around these dimly glowing spheres, some forming in as little as 1 million years.[43] Here, too, tidal stresses around planets or moons might supply enough energy to support life.

Multiple-Star Systems

A large fraction of stars exist as multiple-star systems, usually as binaries, where one star orbits another or both are orbiting a common center of gravity. For a long time it was thought that planets could not exist in such systems, that the tidal pull of different stars would tear a nascent planet apart. More recent work shows that this is not necessarily so. A survey of 131 exoplanets found outside our solar system shows that 23 percent exist in multiple-star systems, usually in orbit around one of the stars in a binary pair.[44] A theoretical study shows that at least 50 percent of the time an Earth-size planet could remain in stable orbit around one of the partners in a binary for at least 4.6 billion years, the current age of our planet.[45] A planet can remain in stable orbit around one star in a binary if its distance is small compared to the separation between the stars, and it can stably orbit both stars if its distance from the center of mass of the two stars is much greater than the separation between the stars.[46]

Disks of material have been found surrounding both stars in a binary, with masses comparable to those surrounding young single stars, suggesting that the process of planet formation could be similar in both cases.[47] Thus, the scene in *Star Wars* of two suns in the sky of the planet Tatooine, the home world of Luke Skywalker, may not be too far-fetched. The frequency of binaries goes down with the size of the star. G-type stars like our sun have stellar companions about 43 percent of the time, but only 30 percent of small M stars exist in binaries.[48]

Galactic Habitable Zones

If planetary systems are common around stars, not all stars necessarily reside in areas conducive to life. The central regions of galaxies are believed to have high levels of radiation that could incapacitate life that tried to get under way there. The core of most galaxies has large, closely spaced stars. Big stars live out their lives quickly, ending in supernova explosions that blast heavy elements into the surrounding medium and forming shock waves that can initiate new waves of star formation.

For a star greater than about 25 solar masses,[49] the residual core may be so dense that it collapses into a black hole whose gravity is sufficient to prevent even light from escaping. Black holes can pull matter and even whole stars into themselves, sending showers of high-energy radiation from their peripheries. Such objects reside in the centers of many older galaxies, including our own Milky Way.[50] The fireworks from black holes and supernova explosions may render the central regions of galaxies inhospitable to life, at least during the most active period of a galaxy's core. One study of the Milky Way indicates that one would have to travel at least 44 percent of the way out from the galactic center to find the habitable zone. Past about 58 percent of the distance out, there may be too few heavy elements to form planets.[51] This galactic habitable zone encompasses roughly 10 percent of the stars in our galaxy, or about 40 billion stars.

But other studies are more optimistic, pointing out that the rate of supernova formation dies down as a galaxy ages. The habitable zone exists in a ring around the central core that expands with time as star burning becomes more peaceful. One study estimates the ring to increase from

10 percent of the radius of the Milky Way when it was 2 billion years old to 28 percent at 4 billion years and 60 percent at 8 billion years. The expanding ring is broadened by both the decrease in high-energy radiation in the center over time and the increase in heavy elements at the periphery as older stars enrich the surrounding space with their remnants.[52]

Star formation is believed to have peaked in the universe about 5 bya, about the time our solar system formed. Overall, it has been in decline ever since and is now only 10 to 15 percent of the fecundity of the universe's baby boom days. But as recently as 1 bya, it was at about 50 percent of its peak, and in small galaxies, star formation has remained at a fairly steady rate from 10 bya until today.[53]

Young galaxies in the early universe probably had more cold gas available to make stars than is present today. The age of galaxies can be judged by how far red-shifted their light is, and a recent study indicates that galaxies 3.3 billion years old had 44 percent cold gas available for star formation, whereas galaxies 5.5 billion years old had 34 percent. This is three to 10 times higher than is found in today's active spiral galaxies.[54] Galaxies have settled down to more sedate lives of star making, although they are forming stars from a more enriched interstellar medium. Our Milky Way galaxy is estimated to still contain some 10 billion solar masses of hydrogen, about half as atomic hydrogen and half as hydrogen (H_2) molecules.[55] It is producing new stars at the rate of about one to four per year.[56] Since most of the large galaxies in the universe (75 to 85 percent) are either spiral or lenticular—with large bands of cool, dusty gas available for making stars—there are still plenty of new stars to be born. Edge-on pictures of spiral galaxies show dark bands of dust and gas where this material still resides. The Milky Way is expected to continue producing new stars for at least another 50 billion years, more than 3.5 times the current age of the universe.[57]

Elliptical galaxies do not have the long ragged arms of spirals, where active star formation is often under way. They contain older stars and hot gas that resists compression by gravity, so little new star formation is believed to occur.[58] But many ellipticals formed by galaxy collisions in earlier epochs of the universe when star-bearing clouds were closer together.[59] These spectacular encounters are not believed to disrupt many stars because the distance between stars is so large. Galaxies are separated by about 20 times their diameter, so galaxy mergers are not uncommon. But stars within galaxies are separated by about 10,000 times their diameter, so galaxies can pass right through each other without many star collisions.[60] But galaxy mergers do lead to bursts of new star formation as their gas clouds mix and are compressed. Because galaxy collisions took place after the first generation of stars had seeded their clouds with heavier elements, terrestrial planet formation should also have been possible in many elliptical galaxies, although they would generally be in orbit around older stars today.

The image on the cover of this book shows a ring galaxy (AM 0644-741) about 300 million light-years from Earth. It formed when a small intruder galaxy fell through the center of what had been a spiral galaxy. Like a rock thrown into a pond, it sent a compression wave outward that condensed dust and gas into a ring of bright, newborn blue stars.[61] At 150,000 light-years in diameter, this ring is wider than our Milky Way. We are seeing these stars as they were 300 million years ago, a mere 2 percent of the current age of the universe. It shows the residual star-making potential that lies within the dust and gas of many galaxies.

Metallicity

Astronomers use the term *metallicity* to describe elements heavier than hydrogen and helium left over from the Big Bang. The heavier atoms are formed inside stars by fusion and returned to the interstellar medium during their lifetimes. New generations of stars and planetary systems form from this enriched medium.

A large study found that planet-bearing stars have higher metallicity than stars without planets.[62] There is a smooth curve of increase so that as the percentage of heavy elements in a star rises, so does the likelihood that it will have planets. There is also a trend of more multiple-planet systems around stars with higher metallicity, although the numbers are not yet large enough to be statistically significant. Only one of 22 stars (4.5 percent) with metallicity below that of our sun has multiple-planet systems, but 13 of 98 stars (13 percent) have more than one planet at metallicity above the solar level.[63]

As a star ages, it continually blows particles into its surroundings via its stellar wind. This culminates in a spectacular expansion at the end of its life, when it will return most of its material to space. For stars less than two solar masses, this is a somewhat gentle process in which the star becomes a nebula that will carry off about half of its mass to the interstellar medium. For stars greater than eight solar masses, this process culminates in a spectacular supernova that will blast most of the star's innards into space. These stars fuse elements together up through iron during their lifetimes, but the force of their cataclysmic explosion creates heavier elements beyond iron, up through uranium and beyond.[64] Thus, as a galaxy ages, its interstellar dust and gas will become enriched with heavier elements. We find that ancient globular clusters of stars that surround the Milky Way have much lower metallicity than stars forming today in the galaxy.[65] Our sun formed about 5 bya and incorporated elements enriched about 1.6 percent above hydrogen and helium. The latest stars forming now incorporate dust and gas at about 2 to 3 percent metallicity.[66] It is some 13.7 billion years since the Big Bang, and still the interstellar medium is enriched to a maximum of only about 3 percent.

A planet was found recently around a star with only 1 percent of the sun's metallicity. This shows that planets were likely possible very early in the history of the universe, when the interstellar medium had been only slightly enriched by the products of stellar fusion. This star is very likely a remnant of a satellite galaxy that merged with the Milky Way billions of years ago, and therefore its planet is the first one we have seen of extragalactic origin.[67]

The possibilities for the existence of rocky planets like our own and life should grow with increasing metallicity. Larger amounts of silicates and metals should raise the likelihood of forming terrestrial planets,[68] and life as we know it depends on elements synthesized inside stars, such as carbon and oxygen, which are the most abundant elements in the interstellar medium after hydrogen and helium.[69] Successive generations of stars provide increasing amounts of these materials to later generations of stars, making solar systems with rocky planets and metal cores more likely over time.[70]

How Rare Is Earth?

If planets and habitable zones are likely all over the universe, what are the chances of a rocky Earth-like planet where complex life could evolve? In 2000, paleontologist Peter Ward and astronomer Donald Brownlee published the book *Rare Earth*, arguing that conditions for complex life on a planet like Earth must be exceedingly rare in the universe. They maintained that we exist at a special conjunction of circumstances that are unlikely to occur often in a galaxy. Their book carried much influence, and their arguments warrant a reply. In large part, this is provided in a more recent book by NASA scientist James Kasting in *How to Find a Habitable Planet*. What is presented next is in large part a duel between these two books. It should be noted, however, that Ward and Brownlee believe that life is probably widespread in the universe, but that it does not often have the conditions to progress to more complex forms. Because they deny the ubiquity of complex life, I will label theirs the *con* position.

Con: A large planet like Jupiter is necessary as a guardian to smaller planets in an inner solar system. Jupiter deflects comets and asteroids with its enormous gravity and thus protects Earth from most life-destroying impacts, such as the asteroid that killed off the dinosaurs. The role of Jupiter as a vacuum cleaner that sweeps up debris headed for the inner solar system was witnessed in July 1994, when it absorbed multiple hits from comet Shoemaker-Levy 9.

Pro: The presence of Jupiter is a mixed blessing. The asteroid belt that exists between Mars and Jupiter is debris from a failed planet that could not form in that orbit because of the strong gravitational pull of Jupiter. Mars's small size is also likely due to the drawing down of material by Jupiter's massive attraction. Were Jupiter not present, there might have been habitable planets at both the orbits of Mars and the current asteroid belt.[71] Asteroids in the current belt still collide and sometimes are swung by Jupiter's gravity into orbits headed toward Earth.[72] A nearby giant planet is, therefore, very much a two-edged sword, and it can pose as a threat as well as a shield to worlds of an inner solar system.

Con: A terrestrial planet needs a large moon to keep the tilt of its spin axis from wandering. The proximity of the Earth's large moon keeps variation of the planet's tilt within 2.5 degrees over 41,000 years, and this allows our climate to remain stable. Among the rocky planets, only the Earth has a single large moon to stabilize it in this way. Mars, for example, with its two small moons, varies as much as 45 degrees in its tilt axis.[73]

Pro: A new theoretical study, more comprehensive than earlier ones, shows that Earth-Moon associations could be fairly common. Their simulations result in 1 in 12 terrestrial planets hosting a large moon like our own.[74]

Even without a shepherd moon, if an Earth-like planet varied its tilt as drastically as does Mars, this would likely occur over periods of at least tens of thousands of years, and this is probably something life on the planet could adjust to.[75] Earth has oscillated between warm periods and ice ages over such time intervals, and this did not prevent complex life, such as herds of mammals, from flourishing over the most recent 3 million years. Climate changes are also moderated in oceans because of the high heat capacity of water, and complex life can evolve in that medium also, as we saw with dolphins.

Finally, variation in a planet's spin axis is not determined simply by its moon or moons. The position of each planet, as well as the pull of its central star, determines the stability of its spin angle. Moreover, the speed of a planet's rotation determines how much bulge it has at its equator, and a bulging waistline affects how much pull the gravity of other bodies can exert on the planet. In sum, one needs to know not only the nature of a planet's moons but also its rotation rate and its position in an entire planetary system to determine whether its spin angle will vary drastically.[76]

Con: Plate tectonics are necessary to keep a planet habitable. They act as a global thermostat that keeps greenhouse gases within tolerable limits. The recycling of rock on a planet's surface has a significant effect on the nature of its atmosphere, as can be seen on our world.

Earth's interior is hot, and molten rock rises whereas cold, dense rock falls in cycles of convection, much as warm air rises and cold air falls in our weather patterns.[77] Earth's surface is broken up into large plates that ride over a convecting mantle of hot rock some 95 km (about 60 miles) below. In some areas, hot magma is rising up to create new land, and in other places, cool surface rock is subducting down under an adjoining plate.

This conveyor belt takes carbon dioxide out of the atmosphere. Carbon dioxide combines with water to form carbonic acid. This acid weathers silicate rocks and forms calcium carbonate (limestone), much of which is deposited on the ocean floor.[78] Carbonate rocks contain some 170,000 times as much carbon dioxide as is now present in our atmosphere.[79] Carbon dioxide is a greenhouse gas, and without this mechanism, our planet would probably be hellishly hot, as is the case on Venus. On the other hand, all the carbon dioxide would be absorbed were it not for the conveyor belt of rock that subducts under neighboring plates. Where it meets the hot interior, it melts and releases carbon dioxide to the air again via volcanic eruptions.

Plate tectonics regulate global carbon dioxide levels through another mechanism. If the planet warms, more water evaporates and falls as rain.

Chemical weathering of rocks increases, making more silicates available to react with carbon dioxide. This process takes the greenhouse gas out of the atmosphere and results in cooling. On the other hand, if a planet cools, weathering decreases, less carbon dioxide is taken out of the air, and warming occurs.[80] Thus, like a thermostat, these cycles have a feedback mechanism that counters trends to make the planet too hot or too cold.

Both Venus and Mars lack plate tectonics, so this mechanism may be rare on rocky worlds. Also, water is necessary to lubricate the movements of plates under one another, so such movements require a planet with oceans.

Pro: Venus is a little too near the sun to orbit in its habitable zone, and were it not for that factor, it might easily have acquired Earth-like conditions. Venus most likely began with high levels of carbon dioxide and water, much like Earth, but its proximity to the sun evaporated its oceans. Water vapor is also a greenhouse gas, so a runaway effect took place. High levels of water vapor and carbon dioxide in the air led to further warming, until the oceans entirely evaporated. Without water as a lubricant, plate tectonics could not sequester the carbon dioxide. High in the atmosphere, water dissociated by solar radiation back into hydrogen and oxygen. The light hydrogen gas escaped into outer space, so the presence of water was lost forever. The result is a dry planet with a thick atmosphere and surface temperature of about 460°C (860°F).[81] But the temperature of Venus was probably mild in its early years, when the sun was putting out 30 percent less energy than it does today, and the planet might even have had conditions similar to those of the Earth in the period when life began here.[82]

Mars is within the habitable zone of the sun but is a victim of being a lightweight. It has half the Earth's diameter and one-ninth its mass. It lives in the shadow of Jupiter, whose massive gravity probably drew off material during Mars's formative years and forever doomed its development. It cooled too quickly to sustain volcanism and plate tectonics. It therefore was not able to recycle its carbon dioxide, and its low gravity led to the loss of most of its atmosphere into space.[83] But Mars clearly sustained water in its early years, and perhaps even conditions similar to Earth when microbial life first appeared here. There are seasonal outgassings of methane on Mars that may be the product of subterranean methanogens that still exist, similar to primitive microbes that live on Earth. Methane could also be the result of low-level volcanic activity, although Earth probes have not detected any other gases that would normally be associated with volcanism.[84] There are Earth microbes that could exist under the conditions we now find on Mars. Had an Earth-size planet formed at Mars's orbital distance, it would probably have liquid water and a substantial atmosphere today.

In sum, the ability to sustain plate tectonics appears to be chiefly a result of a planet's size and distance from its star. Calculations show that a planet between 0.5 and 10 Earth masses within the habitable zone of a star should be able to recycle its carbonate and silicate rocks and support plate tectonics.[85]

Con: A planet needs a strong magnetic field to shield it from cosmic rays and the solar wind of its star. Without this, its atmosphere is liable to be sputtered away in a continual draft of charged particles streaming out from its star. A magnetic field depends on a liquid core of metals in a planet's interior rotating at sufficient speed to set up electric currents that generate a magnetic field. Whether the core is molten depends, in part, on how a planet vents its heat. Plate tectonics allow a gradual escape of heat, maintaining a differential between surface and interior that keeps the upper layer of Earth's core molten. Mars, with its smaller size, cooled too quickly and became entirely solid. It has no magnetic shield and has lost most of its atmosphere.[86] Venus lacks plate tectonics and rotates too slowly (once every 243 Earth days) to create the dynamo needed for a magnetic field.

Pro: Although a strong magnetic field is probably helpful, it may not be an absolute necessity for a habitable planet. Venus has retained a thick atmosphere despite lacking a magnetic shield. Venus is much closer to the sun than Mars is, and it has maintained its atmosphere because of its greater mass. Increased gravitation, therefore, can resist the dispersing effects of a solar wind. A thick atmosphere is also the main shield against radiation from outer space. Although a magnetic field does deflect many charged particles from the sun, the most dangerous high-energy cosmic rays are not deflected from our own world. Instead, they encounter molecules of Earth's atmosphere, resulting in showers of much-lower-energy secondary particles near the surface. These account for only about 10 percent of an average person's yearly radiation dosage, less than from medical X-rays.[87] Thus a substantial atmosphere alone may be sufficient to protect a planet from high-energy particles.

Also, magnetic shields may not be rare. Planet size affects the rate at which a planet cools. Kasting estimates that a world from one-third to

10 times Earth's mass should cool gradually enough to maintain a core of convective metals and thus retain its magnetic field.[88] Planets larger than this are likely to become gas giants like Jupiter or Saturn.

In addition, a rotating world with salty oceans could generate a magnetic field in the absence of a metal core. Such an effect is believed to occur on Europa, a moon of Jupiter.[89] As the universe ages and the metallicity of the interstellar medium is enriched by dying stars, the chance of worlds with metal cores should increase.

In sum, for every argument that a special set of conditions is necessary to maintain Earth's long-term habitability, there is a counterargument of at least equal validity. Reviewing the evidence, astronomer Jeffrey Bennett thinks the requirements for a habitable world may be fairly simple: assemble a planet from rock and metal within a star's habitable zone and be sure it is at least minimally near Earth's size. Such a world has a good chance of having liquid water, volcanism to release gases to make an atmosphere, plate tectonics, a molten core, and fast-enough rotation to produce a magnetic field. "In other words," states Bennett, "any planet born close to Earth-*size* within its star's habitable zone might be expected to be Earth-*like* as well." This leads to the astonishing conclusion that we should "expect Earth-like planets to be quite common in the universe."[90]

Planetary Incubators

Kasting defines the inner edge of the habitable zone for a planet as the place where radiance from its star is sufficient to evaporate water and dissociate it in the upper atmosphere into hydrogen and oxygen (photodissociation). The light hydrogen molecules escape into space and gradually rob the planet of all its water.[91] In our solar system Kasting places this inner limit at 0.95 AU (95 percent of the distance from the Earth to the sun). The outer limit of the habitable zone is where water would be frozen solid. But one needs to take account of the warming effects of greenhouse gases. Based solely on our distance from the sun and the amount of sunlight the Earth receives, it should be frozen solid at −16°C (3°F).[92] Greenhouse gases like carbon dioxide, methane, and water vapor trap solar heat and retain it for the planet. Together with plate tectonics, a natural thermostat can be established that maintains habitable conditions. Kasting places the outer limit of our habitable zone at 1.65 AU. Thus, the habitable zone extends over 0.7 AU (0.95 to 1.65 AU), or about 65 million miles. Kasting points out that the average distance between the four terrestrial planets in our solar system is about 0.35 AU, so we should expect two planets to be able to harbor life. Indeed, we find that Mars is a candidate for life, and that were it a little larger, Mars might have conditions that resemble what we had on the early Earth.[93] If there is anything typical about the spacing of terrestrial planets in our solar system, this indicates that in other planetary systems, at least one planet should be found in the habitable zone around a star.

Stars do not burn uniformly from the time they enter the main sequence of their lifetimes. They gradually heat up as they use up their fuel. The sun at its birth was about 30 percent fainter than it is now, and its luminosity has been increasing about 1 percent every 100 million years.[94] This may make the continuously habitable zone narrower than the habitable zone calculated here. For example, the outer edge of a habitable zone may need to be closer to a star during its fainter early days.

But we should also remember that life on Earth had a very long incubation stage as single cells and did not undertake complex multicellular forms until the last 15 percent of the history of life on our planet (the most recent 600 million years). The earliest life forms were likely hyperthermophiles, some of which today thrive at temperatures above the boiling point of water. Microfossils of cells go back at least 3.4 billion years,[95] to a time when Earth was experiencing 15 times the number of asteroid impacts that it has today, and this clearly did not exterminate the microbial life then present.[96] Life may well extend back at least 3.8 bya, during the period of Heavy Bombardment, when the planet was even hotter and more inhospitable. Our world has also seen repeated glaciations, some severe enough to create a "snowball Earth," when the oceans were entirely, or mainly, frozen over. Yet complex single-celled life was able to ride out these calamities. A study of microfossils from between 750 and 700 million years ago found that both prokaryotic and eukaryotic communities thrived before and during a severe glaciation.[97] These abilities persist into the present. A study of Arctic sea ice in winter, at temperatures from −2 to −20°C (28 to −4°F), found species of bacteria and archaea that could continue metabolizing down to −20°C (−4°F). Even the coldest ice had liquid inclusions with high salt content where these cells could continue to thrive.[98]

Thus planets or moons at the edges of what we consider a habitable zone might exist for a long time with microbial communities. They could act as incubators where cells gradually accumulated the skills that allow more complex associations. The Cambrian explosion on Earth happened after a severe cold spell, and after a period of some 3 billion years during which single-celled life slowly accumulated the genetic programs that allowed it to undertake complex multicellular body plans. The trajectory from jellyfish to human beings is about 600 million years, only about 15 percent of the full history of life on our planet. Complex forms could await the time when a warming sun brings temperate conditions or when accumulating oxygen in an atmosphere facilitates high-energy metabolism. Moons of outer planets, such as Europa, that are believed to harbor liquid water, could, in this scenario, incubate life as single cells for a long time until warmth arrived from its aging star. Planets (and their moons) may even change their positions when a star enters its red giant phase[99] and be moved into orbits with enough warmth to facilitate the transition from simple cells to complex life forms.

A single-species ecosystem was discovered deep underground on Earth in 2008. The bacterium *Candidatus Desulforudis audaxviator* lives 2.8 kilometers (1.7 miles) below the surface in a South African mine, obtaining energy from radioactivity (which dissociates water and supplies high-energy hydrogen gas), and is able to synthesize all the amino acids it needs from inorganic sources. It is thus independent of the sun and all other life forms. Its simple genome contains only about 2,160 genes.[100] Such a species might lie underground on a planet or moon for a long time until surface conditions became more habitable and it became the progenitor of many new life forms.

Life's Toughness

Life can endure much broader conditions than previously thought, increasing its possibility on extrasolar worlds. Besides the temperature tolerances already described here, it can also endure extremes of acidity, pressure, radiation, dryness, and salinity.[101] One species of microbe lives in acid mine drainage at levels of acidity 10 million times greater than the human body can tolerate (pH 0). The Mariana Trench has the world's deepest sea floor, yet it harbors a variety of organisms. Most of these "extremophile" organisms are bacteria, but examples are found in all domains of life, including multicellular animals. The tardigrade, a small invertebrate, can, in its dormant state, survive temperatures from −253 to 151°C (−487 to 304°F), as well as a vacuum, high pressure, and radiation thousands of times greater than what would kill a human.[102] Antarctica has some of the coldest, driest areas on Earth, yet a variety of bacteria, lichens, and algae live there.

The nematode *Halicephalobus mephisto* is a small roundworm only about 0.5 millimeter (0.02 inch) long, but it has digestive, nervous, and reproductive systems. It has been found living in South African gold mines deep underground—at depths of 0.9 to 3.6 kilometers (0.56 to 2.2 miles), in high-temperature and low-oxygen conditions—feeding on bacteria.[103] The finding of multicellular life so deep underground prompted the director of NASA's Astrobiology Institute, Carl Pilcher, to say, "It is entirely plausible, in fact extremely likely, that subsurface environments like those described in these papers exist on other worlds in this solar system and in other planetary systems. We can now say that worlds with such subsurface environments could, in theory, harbor subsurface life, both microbial and multicellular."[104]

On the basis of differences in DNA sequences, structure, and biochemistry, biologists divide life into three domains: bacteria, eukarya, and archaea. At the base of each lineage, nearest the common ancestor of all life, are hyperthermophiles, species able to thrive at high temperatures. Most grow optimally above 80°C (176°F), and some prefer temperatures above boiling water.[105] They are probably similar to the earliest cells that evolved while Earth was still hot and still subject to the heavy bombardment of asteroids that formed the solar system. Hyperthermophiles generally are also very tough survivors, able to withstand freezing down to −140°C (−220°F). They indicate that early species on our planet were more like street fighters than hothouse plants, able to endure extreme conditions and change.

Life also has the ability to endure higher levels of radiation than we normally experience on Earth. A bacterium that can survive high radiation, *Deinococcus radiodurans*, has evolved unique mechanisms to repair DNA damage.[106] This cell can survive radiation doses 3,000 times higher than what would kill a human. Its repair enzymes can reassemble its chromosome after it has been fragmented by radiation and can also repair hundreds of double-stranded breaks in its DNA. It has been found growing in the cooling water of nuclear power plants, and it also has unusual resistance to high levels of ultraviolet light. It is not alone in these abilities. A bacterium found in the Sahara desert, *Deinococcus deserti*, has equal resistance to gamma radiation as *D. radiodurans*, and even higher resistance to ultraviolet light. Six other species of bacteria have been shown to survive radiation by gamma rays from 25 to 1,600 times the dose that would kill a human.[107] These species suggest that life might tolerate radiation levels on a planet that is closer to its star, or has less shielding by its atmosphere or magnetic field, than we currently have here on Earth.

It is not just simple cells that can flourish in extreme conditions, but also animals with complex nervous systems. Lake Natron, in Africa's Great Rift Valley, is located in northern Tanzania. Its shallow waters are among the most alkaline of any lake on Earth, with a pH between 9 and 10.5 (near that of household ammonia), and temperatures in its waters can reach 50°C (120°F).[108] Nevertheless it is a favored feeding and breeding ground for the lesser flamingo (*Phoeniconaias minor*) and contains an endemic species of fish, the alkaline tilapia (*Oreochromis alcalicus*). The birds feed almost entirely on a species of blue-green algae that thrive in these highly saline waters. They filter-feed on these cyanobacteria with their beaks, separating them from water that would be toxic to most animals.[109] Flamingos have a complex social life, breeding in flocks of hundreds of thousands, yet pairs are monogamous. Both parents incubate the egg and must continue feeding the young for about 70 days after hatching.[110]

An even more astounding feat of survival is achieved by Emperor penguins (*Aptenodytes forsteri*), which breed in Antarctica during the coldest time of the year. The female leaves her single egg with the male, who balances it on his feet and endures temperatures as low as −62°C (−80°F) and winds up to 180 km (112 miles) an hour in the perpetual night of Antarctic winter.[111] The males huddle in a large shuffling crowd to shelter themselves from the harsh winds and take turns in being exposed at the outermost layers. Their epic endurance is portrayed in the Academy Award–winning film *March of the Penguins*. They do not eat for four months, and by the time the female returns to relieve him, the male will have lost up to half his body weight. He must then trek some 97 km (60 miles) over the ice to reach open water before he can feed again.[112]

These feats of survival belie the idea that a planet must have long-term temperate conditions for complex life to survive. On another world that was colder or more alkaline or acidic than our own, species such as these might be at the base of a diversification that occupied many different niches. We would be the extremophiles on such a world, our planet's average temperature or pH seeming scarcely tolerable.

Finding Life

Let us ignore for the moment that life might be possible around a variety of stars and on moons as well as on planets. With a conservative estimate, how widespread might conditions for life be in the universe? Astronomer Peter Ulmschneider excludes all stars that live less than 5 billion years and also small stars that would have planets in tidally locked orbits, leaving only sunlike G stars.[113] He also excludes stars that are not high in metallicity and those that exist in multiple-star systems, leaving one in 40,000 stars with planets suitable for life. At current estimates of the Milky Way containing about 400 billion stars,[114] that comes to 10 million stars in our galaxy. Kasting also rules out stars that are not sunlike or that exist in multiple-star systems and arrives at 4 billion Earth-like planets in our galaxy.[115] J. P. Guo calculates 7.6 billion habitable planets around G stars, but when M, K, and F stars that might also have habitable zones are included, the number of habitable planets in the Milky Way reaches 45.5 billion.[116]

Taking the middle value of 4 billion planets and multiplying by the estimated 100 billion galaxies in the universe, we reach 400 billion billion planets that could harbor life in the cosmos. That does not mean, of course, that these worlds actually have life or have developed advanced civilizations that could communicate with each other.

Fermi's Paradox

If life is indeed widespread across the universe, where is everybody? It has been estimated that if advanced life arose on another world billions of years ago, it should have been able to spread across the galaxy by now. This question was posed by the famous physicist Enrico Fermi at a seminar for fellow scientists of the Manhattan Project, who were working on the first atomic bomb. Many immigrant scientists, refugees from Hitler's Europe, worked on this project, including Hungarians, who wore strange clothes and spoke a language incomprehensible to the others. One of them, the physicist Leo Slizard, replied that the aliens "are among us, but they call themselves Hungarians."[117]

If there are many habitable worlds in our galaxy, we may be overestimating the ease of contact. At the speed of the fastest space probes we now have, roughly 50 kilometers (31 miles) per second, it would take about

25,000 years just to reach the nearest sunlike star, Alpha Centauri.[118] Perhaps there will be new sources of energy to power ships in the future, like the matter and antimatter engines of *Star Trek*. Calculations show that to accelerate a ship the size of *Star Trek*'s *Enterprise* to half the speed of light would require 2,000 times as much energy as the entire Earth presently uses in a year.[119] Also, as a ship approached the speed of light, atoms and small particles adrift in space would hit it with deadly force.

The problems of signaling other worlds may be no less daunting. Our galaxy is 100,000 light-years across. If advanced civilizations were separated by just 1 percent of this distance, it would take a thousand years for a radio or television signal from Earth to reach another world, and another thousand years for an answer (one theory holds that ET has seen the content of our television programs and that is precisely why he decided not to contact us).[120]

Planets with advanced civilizations may be separated by huge spaces of time as well as distance. If a technological civilization lasts a million years, and they have been possible since the second generation of stars (about 13 bya), they endure for one-hundredth of 1 percent of that time span. The flowering of one planet may happen at a very different time from that of another, like candles that flicker on and off at widely separated times and places. Advanced technology might also sometimes lead to self-destruction.

But I would like to suggest another possibility for the apparent silence of the heavens. Perhaps, after an infatuation with technological progress that lasts from hundreds to thousands of years, most civilizations find that inner moral and spiritual development is much more important. They become more interested in exploring the universe spiritually than by mechanical means. This may seem outlandish in our technological society, obsessed as we are with the latest electronic gadgets, but there have been periods in history when people sought out the best spiritual teachers and focused their minds on inner development. They were more concerned with growing emotionally and spiritually as people than building outward physical structures. Such was the time of Buddha in Asia, for example, and those ideas occupied the best minds of their society longer than has a scientific culture in our own. Thus, between civilizations that destroy themselves and those that focus on inner development, the period of mechanical exploration may be relatively brief and separated by large amounts of time and distance between planets. The universe could be teeming with life, yet we find the heavens silent with respect to their signals or starships. Given the history of the clash of civilizations on our own world, that might be a good thing.

Indicators of Life

Theoretically, one would not need to travel to another planet to find strong indicators of life. One could tell by the light given off from its atmosphere.

Earth reflects about 30 percent of the sunlight that strikes it. Planets also radiate heat at thermal infrared wavelengths.[121] Gases in the atmosphere absorb distinctive wavelengths of light, and these can be identified when viewing the spectrum of light given off by a planet. Oxygen (O_2) has three distinct bands, the most prominent at 0.76 micron (visible light extends from 0.4 to 0.7 micron). Ozone (O_3) is formed in the upper atmosphere from oxygen by sunlight, and it has a prominent absorption band at 9.6 microns. It is a sensitive indicator of oxygen below, although even low levels of oxygen can produce ozone.[122] Water, methane, and nitrous oxide (N_2O) each have distinctive absorption bands also, and these were detected from space by NASA's *Galileo* spacecraft as it looked back on Earth on its voyage to Jupiter.[123]

Measurements like these are strong indicators of the presence of life. Oxygen is a highly reactive atom and normally would be bound up with other elements in the heat of planet formation. It was gradually liberated on Earth by photosynthesis. But the presence of reduced gases like methane (CH_4) or nitrous oxide would be further indicators of biological activity. Normally, these would be taken out of the atmosphere by reaction with oxygen, but they are continually replenished by the metabolism of cells like methanogens. Thus the presence of substantial levels of reducing gases (like methane) and oxidizing gases (like oxygen) at the same time on a planet would be a strong indicator of the activities of life.[124]

The main problem is seeing light from planets next to their much brighter stars. We have just begun to obtain spectra of planets by the transit method. One technique measures the combined spectra of planet plus star when the planet is in front of the star, and the spectrum of the star alone when the planet is behind it. By subtracting the contribution of the star, one can obtain a spectrum for the planet.[125] The first rough measurements of gases in the atmospheres of giant planets have been obtained in this way. But there are exciting prospects for the near future. NASA's James Webb Space Telescope is scheduled for launch in 2016, and its mirror for infrared measurements, where these spectrum contrasts are best obtained, is seven-and-a-half times larger than the existing infrared space telescope, the Spitzer.[126] Moreover, we have just begun to apply these techniques to dim red M stars, which are the most numerous in our galaxy. These stars have a diameter of only 10 to 30 percent of the sun, and a planet the size of the Earth would cover a measurable fraction of its surface, about 1 percent. The light from these dim stars is less likely to overwhelm that of a planetary companion. As Kasting states, "It is thus conceivable that potentially habitable planets could be identified and perhaps even characterized within the next 10 years."[127]

Detecting Planets

Fortunately, finding planets outside our solar system is a much less daunting task. The first extrasolar planet was discovered in 1995, and the pace of discovery has been accelerating in every year since. In 2006, the tally was 205, and the count in June 2011 was 555.[128] Until recently, the main way to find extrasolar planets was the radial velocity method, which measures the wobble of a star as a planet orbits around it, pulling it to and fro by gravitational interaction. In fact, we measure not the position of the star directly, but the shift in wavelength of its light. If the star is pulled toward us, its wavelengths become slightly shorter, toward the blue end of the spectrum. If it is pulled away, its wavelengths become slightly longer, toward the red. A similar Doppler shift happens to sound: the horn of a car speeding toward us has a higher pitch (shorter wavelength), and it sounds lower after it passes and speeds away (wavelength stretched out). Because most planets are a small fraction of a star's mass, the amount of wobble they induce is small, and the alteration in the wavelength of light is difficult to detect. The radial velocity method, therefore, has inborn biases, and it should not be surprising that a large fraction of the planets it has found so far are Jupiter-size and close to their star. Near orbits will induce more frequent wobbles than distant ones that take years to complete. Also, the planet must be passing at least partly within the line of sight between the observer on Earth and the star to shift the light back and forth. An orbit that made the star wobble up and down, for example, with respect to the observer would not be Doppler shifted in a way that we could measure. Thus, this method must miss many planetary orbital inclinations that are not favorable for Earthbound viewing.

A theoretical study indicates that we are only seeing the tip of the iceberg of what is really out there. It predicts the existence of a large number of smaller planets in orbits around stars at distances between a half and seven times the distance from Earth to the sun (0.5 to 7 AU). The study estimates that we are so far seeing only about 9 percent of all the existing planets and that "almost all stars with no (giant) planet detectable today should harbor low mass planets."[129]

In March 2009, NASA successfully launched the Kepler space telescope, which is capable of monitoring 156,000 stars in a small section of the Milky Way near the Northern Cross. Using a 95-megapixel camera, it measures the dimming of a star as a planet passes over its face. It can detect a $\frac{1}{10,000}$ dip in brightness when an Earth-size planet moves in front of a sunsize star over a 12-hour period.[130] In February 2011, results of the analysis from the first four months of data collecting were announced. The telescope found 1,235 candidate planets, more than tripling the number of exoplanets identified up until that time. Sixty-eight were Earth-size planets, and 288 were super-Earths, with a radius of 1.25 to two times Earth's radius. Fifty-four were in habitable zones, and five were both Earth-size and orbiting in habitable zones.[131] Many of these sightings still need to be confirmed by Earth-based telescopes, but the lead investigator is convinced that most will be verified.

Until this time, most exoplanets had been identified by the Doppler shift method already described here. Almost all of them appeared to be larger than Jupiter and in close orbit around their stars, so it seemed that Earth-like planets might be rare.[132] But some researchers pointed out that this may be an instrumentation problem, with star wobbles much more likely to be seen when their planetary companion is large and nearby. Kepler has stood this initial estimate on its head. Its telescope finds that only 15 percent are Jupiter-size or larger, while 29 percent are between one and two times Earth-size. The biases may not end there. Earth takes a year to orbit the sun, and if Kepler were viewing a solar system identical to our own around a distant star, four months of data probably would not be enough to reveal the existence of an Earth-like planet. Moreover, a planet must pass between a star and a distant viewer to occlude its star's light. There will be many orbital inclinations that Kepler misses. Finally, the space telescope's field of view covers only $\frac{1}{400}$ of the sky, and Kepler is near the edge of sensitivity for catching the presence of small, Earth-size planets, so there may be many more riches out there.

Says William Borucki, NASA's principal science investigator for the Kepler mission, "The fact that we've found so many planet candidates in such a tiny fraction of the sky suggests there are countless planets orbiting stars like our sun in our galaxy."[133] Extrapolating from the findings so far leads to an estimate of 20,000 habitable-zone planets within 3,000 light-years from Earth, and hundreds of millions of such planets in the Milky Way Galaxy.[134] Years of verification lie ahead, as well as data about planets in longer-term orbits, but these results support the theoretical studies described earlier. They may be among the most significant findings of modern science and they have profound implications for the place of life and humanity in the cosmos.

12

The Apex of Nature

A star is an accident. It takes a certain volume of dust and gas, traveling at the right speed and density, to begin the gravitational collapse that will ignite the nuclear fires of a newborn star. But conditions after the Big Bang were such that materials for this accident were provided all over the universe, and the sky above us is studded with billions of points of light. It may be further asked whether solar systems, habitable planets, life, and intelligence are not also "accidents waiting to happen."

As we saw in chapter 11, planet formation is probably an integral part of star formation. Unless an interstellar cloud condenses with perfect symmetry, there will be material left over that stabilizes in a ring around the equator of a nascent star, and from this material a set of planets is likely to be born. At least one of the planets in that solar system has a good chance of being in the habitable zone of that star. Ninety-six percent of stars are of two solar masses or less[1] and therefore should shine long enough for the evolution of advanced life, judging by the time it has taken on this planet. Life may be inevitable wherever conditions are provided for its appearance. Carbon and oxygen are two of the most abundant elements seeded into space by stellar burning, and they are ideally suited to creating complex organic molecules and water (H_2O), two essential components of life as we know it. Precursors for all the building blocks of life are already forming in interstellar space and raining down on us in carbonaceous meteorites, as they are likely to do on early worlds everywhere. We have now found nonbiological pathways to all the components of cells, including amino acids, sugars, lipids, and RNA. To be sure, we do not yet know how these components were assembled into the first cells, but plausible scenarios can be written for energy sources on the early Earth, such as black smokers found underwater. The early appearance of life on our planet, almost from the time when it had cooled sufficiently to allow life, argues for a natural process that happens wherever conditions permit it.

In this book, I have viewed life in terms of two paradigm strategies. All life must stay adjusted to change, and it can do so with either high- or low-information pathways. It can produce many individuals with short genetic programs that develop quickly, selecting those few that are best suited to meet current conditions, or it produces individuals with high-information content in genes, brains, or both that have wide behavioral versatility to respond to change. The latter are created in smaller numbers and tend to entail more parental investment per individual than do low-information organisms. I discussed nervous systems as an extension of the information gathering of genes and showed how brains have gone beyond the genes in potential information content as measured in bits, as Carl Sagan suggested years ago.[2] This can be seen in the open-ended development of brain synapses, where experience, in part, determines which neurons will be wired together. Humans have extended this strategy further by gathering information in languages, books, and computers, and our dominance on the planet indicates the overwhelming success of this strategy in dealing with the challenges of nature. Each level builds on the previous ones, increasing its information-gathering capacity and the speed of response. Human culture is not a bizarre offshoot of a language-specialized primate but an extension of a fundamental strategy life has used from its beginnings in seeking homeostasis, maintaining a steady environment conducive to the flourishing of life.

I described the advanced cognition of four animal groups that have large brains on either an absolute or relative scale. Relative brain size is calculated in terms of EQ, encephalization quotient, which measures how much larger an animal's brain is than that of an average animal of its body size. This recognizes animals with small bodies, but high brain:body ratios, such as corvids. Absolute brain size refers to the full weight of a brain, which will be high in an animal with large body size, such as an elephant. As brains become larger in absolute terms, they appear to devote a smaller fraction of total brain neurons to mere physiological control of the body.[3] Large brains, in either an absolute or relative sense, appear to have many "extra neurons" that can be devoted to higher cognitive functions. The four animal groups portrayed came from very different lineages and ecologies, yet they share many qualities in common, such as complex communication and social organization, tool use, imitation, insight learning, and mirror self-recognition. I described this overall as an emergent self, able to view itself, conspecifics, and the environment with higher objectivity. I suggested that a large brain may have extra circuits that are able to abstract patterns from lower circuits to create a higher "third-order" objectivity.

In this sense, one may say that an apex of the evolutionary process is a person. The summit of the information-gathering process is not a mind that acts like a computer. It is an emergent self with much to express and rich relationships with other individuals. It recognizes and values other individuals beyond the bonds of just genetic relatedness. Part of this self-expression is shown in the dexterity of its appendages and the ability to manipulate its environment in intricate ways. Higher objectivity is also shown in insight learning and the ability to invent new tools. Imitation may be the result of being able to form an image of the other and then seeking to match it with one's own actions. Higher objectivity is also indicated by these four species being among the few animals that show evidence of self-recognition in mirror experiments.

The high-information pathway has come to fruition in the primate order on our world, but it is emerging in other lineages also and might take very unexpected forms on other planets. Rather than reflecting rampant anthropomorphism, this view suggests that there is a complex of characteristics that are involved with each other and that they have emerged repeatedly in the evolutionary process because they provide a distinct advantage. Primates were probably predisposed to show an advanced form of these qualities on our planet because for a long time they have had high brain:body ratios compared to those of other species, and because their experience in trees provided unusual dexterity with hands, but this suite of qualities could also emerge in other lineages with a very different evolutionary history.

In suggesting that stars, planets, and life are "accidents waiting to happen," I do not mean to suggest some kind of strict determinism. The universe could have a destiny without being deterministic. If you throw a rock down a mountainside, it will be subject to chance events, and each time will probably land in a different place. Nevertheless, its destiny will always be in a particular direction. The high-information pathway, I suggest, is integral to the evolutionary process, but on another world it may come to fruition in a species that looks very different from our own.

The destiny of our universe—and its potential for stars, planets, and life—were probably set within the first microsecond of its existence. But it also had room for contingency, so that the forms that evolve could differ greatly on different worlds.

The Improbable Universe

It may be that when the universe was only 10^{-35} second old, at the end of an inflationary period that followed the Big Bang, that all the conditions were set for the future evolution of life. The early universe had just the right amount of lumpiness, an excess of matter over antimatter, and just the right relationship among the four basic forces of physics to allow the eventual appearance of stars, planets, and life.[4] This fine-tuning includes a variety of improbable relationships, as described in physicist Michael Mallary's *Our Improbable Universe*:

1. The early universe was very smooth, but quantum fluctuations created just enough lumpiness—on the order of 50 parts per million—to allow gas to condense into galaxies in the future universe.[5] Had the fluctuations been greater, there would have been an excess of black holes in the cosmos; had it been much less, few stars would have been able to form.[6]

2. Theory dictates that for each particle of matter created at the Big Bang, there should be a particle of antimatter. If the two meet, they annihilate in a burst of pure energy. But there was a slight imbalance at the beginning of the universe—an asymmetry of parts per million that current theory cannot explain—that allowed a slight excess of matter.[7] For every billion particles of antimatter, there were a billion and one particles of matter, and all the visible cosmos around us formed from that slight excess. Without it, matter and antimatter would have mutually annihilated each other, and the Big Bang would have ended as an expanding ball of light.

3. There was a precise balance between the mass-energy of the early universe and its rate of expansion. If density had been a little greater, the Big Bang would have been quickly followed by a Big Crunch as gravity collapsed the universe again. Had the energy of the explosion been a little greater, it would have resulted in a diffuse gas that could not have clumped into galaxies and solar systems. The ratio between the mass-energy of the

early universe and its rate of expansion had to be precise to within one part in a trillion trillion trillion trillion trillion (10^{60}) to produce the cosmos we know.[8]

4. The strong force that holds the nucleus of atoms together is just the right strength to permit long-lived stars. If it were 0.5 percent stronger, hydrogen fusion would proceed too quickly, and all stars would live less than 100 million years. If the strong force were a few percent weaker, fusion would occur only inside giant short-lived stars. Neither scenario would provide the long time periods that the evolution of life seems to require.[9]

5. Other fundamental forces of nature also appear to be finely tuned. If the weak force were somewhat weaker, much less hydrogen would have formed in the early universe, and hydrogen is the main fuel for stars.[10] If gravity were stronger, stars would burn through their fuel faster and die younger.

6. Empty space is filled with a mysterious force known as *dark energy*. Its value is 10^{120} times smaller than the quantity predicted by quantum mechanics. If it were 119 powers of 10 less, rather than 120, there would be too much repulsive force in the universe for galaxies, stars, and planets to form. Physicist and cosmologist Paul Davies calls this the "biggest fix" of all in the parameters that define our universe, the equivalent of flipping a coin and getting 400 heads in a row.[11]

The anomalies also appear in a variety of basic particles in physics and the forces between them. If protons were 0.2 percent heavier, they could decay into neutrons, and atoms would be unstable. If the electromagnetic force were 4 percent weaker, there would be no hydrogen or long-burning stars. If the weak force were much stronger or weaker, there would be no hydrogen, and supernovas would not seed space with the heavy elements they synthesize in their interiors.[12]

The Fitness of Carbon and Water

The physics of fusion seem tuned peculiarly to produce large quantities of carbon and oxygen as byproducts of star burning, and studies indicate that these two elements will likely be important for life wherever it may be found. Why these two atoms more than others on the Periodic Table of Elements?

Both carbon and water are suited ideally for the requirements of life—water as an ideal solvent, and carbon for its ability to form large, complex molecules. Water is formed readily when oxygen is added to the abundant hydrogen left over from the Big Bang. Chemical pathways to water formation have been found both in the cold of outer space and the warm neighborhood of protostars.[13] Water ice is believed to be the most abundant solid phase compound in interstellar space across the whole universe.[14]

Carbon is unique among the roughly 90 elements of the Periodic Table that are commonly found in nature in that it can form four strong covalent bonds with itself.[15] It can thus form long chains and also branch out into complex three-dimensional structures. There are close to 10 million known organic compounds (compounds containing carbon), more than 25 times the number of inorganic molecules.[16] As you go farther down the Periodic Table, atoms add extra shells of outer electrons, and their bonds with other atoms become weaker, so they are less suitable as the strong backbone of large molecules like proteins or DNA. Silicon, for example, is just below carbon in the Periodic Table, and like carbon it can form four bonds with other atoms, but the Si-Si bond is too weak to form chains longer than two silicons. Plants take carbon dioxide (CO_2) out of the air to form complex molecules like sugars. Animals (and plants) metabolize these molecules for energy and return carbon dioxide to the air. Silicon-based life would have a much harder time doing this: silicon dioxide (SiO_2), or silica, is a strong solid at room temperature and the main ingredient in sand and glass.

Water is also a unique substance with many qualities suited for life. If there were a planet where gold and water were equally common, water would be the far more precious substance because of its many talents. Cells are 70 to 95 percent water, and humans cannot go without water for more than about a week, whereas we can fast without food for a month or more.[17]

All life that we know of is a complex mix of chemical compounds, most of which have intricate three-dimensional shapes. The molecules float in a solvent that allows rapid exchanges between components. All cells have remained small, in part, because they rely on diffusion to bring molecules together. On land, you need precise tracks to allow A and B to encounter each other at the right times and angles. But in a solvent like water near room temperature, molecules are continually encountering each other at high rates and from many angles. In effect, all things are being tried out, and those that fit (e.g., have surfaces that mesh) stay together, whereas those that do not will quickly separate again. On land, you have to design complex circuit boards to provide a road map for a series of functions. In liquid, you can allow all things to happen, and only ingredients with the right shapes, or at the right concentrations, will proceed together. On land, two things fit if they have compatible shapes. In water, two things stay together if they are compatible in shape, electrically charge surfaces, and have hydrophobic or hydrophilic interactions. In other words, the information capacity for transactions in liquid is much higher than on dry land. Life on Earth began in a liquid, and everywhere still carries around little lakes inside cells to carry on its business.

Different kinds of bond strength also foster exchanges in water. For example, the backbone of DNA is made with strong covalent bonds. The coding part of the molecule, where the rungs of the ladder come together, is held with hydrogen bonds that are about 20 times weaker than covalent bonds. In water, the amount of energy needed to pry open the ladder and make use of the code inside is much less than what would disrupt the backbone, so the molecule stays intact while it is being used.

Water also has a high specific heat, meaning it takes a lot of energy to heat up water, and it retains a lot of energy. This is because the many hydrogen bonds between water molecules connect one to another like so many bedsprings. When you heat up water, you not only speed up individual molecules, you stretch the bedsprings, so it takes extra energy to do so. Water thereby moderates the climate of a planet by absorbing excess energy during warm months and giving it back in winter. For that reason, coastal cities like San Francisco have temperate climate all year round. Water can help make a planet habitable by retaining heat from its sun and giving it back gradually, and also by lubricating plate tectonics, as we saw in chapter 11.

Water is nearly unique among substances in becoming less dense as it becomes a solid. As it approaches freezing, its hydrogen bonds spread out at maximum angles, making water less dense. Ice therefore floats on water instead of forming dense blocks that would come crashing down on life below. A frozen surface also helps insulate lakes and oceans from colder air temperatures above, further protecting marine life. Most substances become denser as they cool. A cold bar of iron is a little shorter and denser than a heated bar. Water is rare in its ability to become less dense as is solidifies, and is also the only substance that can exist as a solid, liquid, or gas within the common range of temperatures on our planet. The polarity of water also makes it an excellent solvent for polar substances, and it forms shells around ions like sodium and chloride (NaCl), keeping them suspended in water as separate ions.

Besides its presence in water, oxygen is also probably an important gas in the atmosphere of any planet with complex life. In metabolism, electrons are passed "downhill" to compounds that hold them ever more tightly as a way to harvest energy, much as falling water can be used to power mechanical work. Passing electrons to oxygen provides the largest free-energy release of any element on the Periodic Table, except for fluorine and chlorine. As astrobiologist David Catling and colleagues point out, of these three gases, only oxygen is stable enough to accumulate in a planet's atmosphere.[18] Compounds of both fluorine and chlorine are highly corrosive and used in bleaches and strong acids. Aerobic metabolism provides about 10 times the energy of anaerobic metabolism (without oxygen), meeting the high-energy demands of complex organisms. Only aerobic organisms grow large on Earth. The brain is one of the most expensive organs metabolically, and all the largest-brained animals are warm-blooded, with high oxygen needs.

Oxygen dissolves in water as well as being stable in atmospheres, facilitating life in both media. Catling and colleagues propose that it was the long time needed to accumulate high concentrations of oxygen in our atmosphere that delayed the proliferation of multicellular life on our planet.[19] This happened during the Cambrian explosion, in the most recent 15 percent of the history of life on our world.

The Cosmic Abundance of Carbon and Oxygen

Fusion processes inside stars seem to be specially tuned to produce large quantities of carbon and oxygen, more of the improbable numbers that define our universe. Besides the primordial hydrogen and helium formed at the Big Bang, they are the two most abundant elements distributed through interstellar space.

The production of carbon 12 depends on the fusion of three helium 4 nuclei. The likelihood of three helium nuclei striking each other inside a star at just the right time and place is small. Two nuclei could hit each other and form beryllium 8, but this is highly unstable and usually disintegrates immediately.[20] However, there is an excited state of beryllium 8 that lasts just long enough for a third helium to join the party so that carbon 12 can form.[21] This brief stable period (a resonance) results in carbon being the second most abundant element produced by stars in the range of one to eight solar masses.[22] The addition of a fourth helium nucleus produces oxygen 16. This could lead to all the carbon being converted into oxygen. But the stable resonant state for oxygen is 1 percent lower than expected for a typical nuclear energy level spacing, so that carbon plus helium fusion has a little too much energy to be stable there.[23] The result is that both elements are produced in a ratio of about 1:2 for ^{12}C:^{16}O.[24] Thus an unexpected excited state of carbon allows it to form abundantly in stars, and the absence of a corresponding state in the oxygen nucleus keeps all the carbon from being converted to oxygen.[25]

These processes take place in the majority of stars for most of their lifetimes, ensuring that carbon and oxygen will be abundant all over the universe. A study indicates that the earliest giant stars after the Big Bang were already probably producing fair amounts of carbon and oxygen.[26] Combined with the ubiquitous hydrogen, this indicates that water is likely plentiful everywhere in the universe. Water was the most abundant condensate in the formation of our own early outer solar system, and simulations of planet formation in habitable zones indicate that a majority of them should receive an inventory of water equal to or greater than that of the Earth.[27]

Entropy and Order

When entropy is discussed, gravitational effects usually are ignored. In many cases, a uniform distribution of particles seems to have the greatest disorder and maximum entropy. For example, if you open a bottle of perfume in a room, you begin with order: all the perfume molecules in the bottle and air in the rest of the room. As the random motions of gases encounter each other, the perfume is gradually wafted around the room, and air infiltrates the bottle. Maximum entropy is attained when air and perfume are equally distributed all over the room as well as inside the bottle.

But when large masses and gravity are considered, nearly the opposite arrangement is true. The universe began with a uniform plasma in which all matter and energy were melded at very high temperature. In the Big Bang, this plasma rapidly expanded and began to cool and clump together. The uniform plasma is considered to have low entropy, and as it clumps together under gravity, entropy increases. The ultimate destruction of order and maximum state of entropy occurs when gravity collapses matter into a black hole.[28]

At the Big Bang, all the matter-energy of the universe began in an extremely tiny space. Why it exploded instead of collapsing into a black hole is unknown. In fact, the force of the explosion was finely tuned with the total matter-energy to allow a long-lived universe that could evolve galaxies, stars, and planets, as already discussed. An enormous potential energy for future activities was thereby provided. Gravitational attraction was stretched out like a rubber band, and as it began to pull matter together into galaxies, that potential energy was turned into the kinetic energy of heat that would begin fusion processes inside stars. The high energy of a sun, in turn, was able to power the buildup of order and life on a planet, as we experience on Earth. The universe is gradually expending the low entropy of the plasma that was present at the Big Bang to the ultimate high entropy of disorder, as the second law of thermodynamics requires. But in the meantime, it can use the enormous capital provided at the universe's origin to build up highly ordered local structures.

Figure 12.1 shows the binding energy of nuclear particles as stars gradually build up the elements of the Periodic Table. Recall that the universe began with about 75 percent hydrogen and 25 percent helium after the Big Bang. As these simple nuclei are fused together, they release energy. Notice that the steepest part of the curve, and therefore the highest potential to release energy, comes from fusing hydrogen together into helium. This is the main power source of the stars.

Here nature again seems to be peculiarly tuned to provide conditions for the future appearance of life. A second well of potential energy becomes available. As gravity releases its potential energy by clumping matter together, it compresses atoms that have a maximum potential to release their energy. Most matter begins its existence as hydrogen, the atom with the highest potential energy. Long-lived stars are possible through hydrogen fusion, and the main byproducts of the process, as already discussed, are carbon and oxygen. Today, 13.7 billion years after this process began, the interstellar medium is enriched only about 3 percent with heavier elements, and trillions of years of potential star burning still lie ahead of us. The cosmos is still young. If the universe were a windup toy, like one of those little cars with a spring that you tighten or a set of chattering teeth, it was set on the table with its coil near maximally

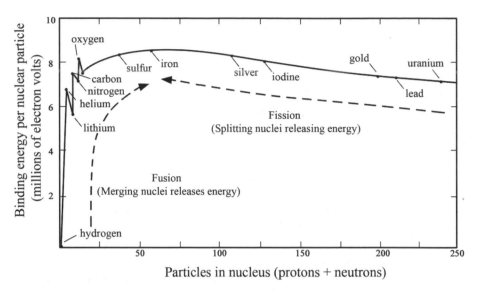

Figure 12.1 The binding energy of nuclear particles (protons and neutrons) and some of the common elements they form.
Fusing particles together releases energy from hydrogen up to iron. Splitting nuclei (fission) of heavy elements like uranium releases energy, but in lesser amounts, as shown by less steepness in the curve.

wound. Today, some 13.7 billion years after the fiery beginning, things are still going on with furious activity, with no sign of a letup.

As figure 12.1 shows, fusion releases energy up through the element iron. At the peak of the curve of binding energy, iron is the ultimate ash of star burning. As it accumulates in the center of large stars, it absorbs energy but cannot release it through further fusion. Energy thus builds up in a large stellar core until it explodes in a supernova. In the heat and pressure of that cataclysmic event, all the elements heavier than iron are formed, up through uranium and beyond. Energy can be released by splitting these atoms (fission), but that potential ends again at iron.

Hierarchies of the Universe

Our universe is believed to have begun with a primordial unity in which its basic forces were joined and all matter and energy were melded into each other. As the universe expanded and cooled, it began to differentiate, first into subatomic particles, then atoms, and then stars that began to build up all the elements of the Periodic Table. All chemical compounds, including the biochemistry of life, derive from those elements. This gathering complexity is depicted by the astrophysicist David Layzer in figure 12.2. The increasing complexity was paid for by the potential energy present in the universe at the Big Bang.

Each level in this hierarchy rests on the binding strength of the previous level. This can be seen by the amount of energy it takes to break up a given structure.[29] Atoms generally lose their electrons, and molecules break up at a few thousand degrees Kelvin (3×10^3 K). Nuclei of atoms break apart into protons and neutrons at about 10 million times higher temperature ($10^{10°}$ K), and these particles split into their constituents at temperatures about a thousand times higher still (10^{13} K). Thus, each level of organization rests on the binding strength of the previous level and makes its existence possible in a hierarchy of increasing complexity.

Life builds most of its stable structures with strong covalent bonds, but many of the exchanges between those parts are done with weaker interactions: hydrogen bonds, ionic bonds, and hydrophobic or hydrophilic contacts.

Layzer's hierarchy of time-bound order shows the gathering complexity of the universe and how each level rests on an earlier one. Like Chaisson's graph of energy rate density (see figure 9.3), it calls for a perspective that stretches from the Big Bang to the evolution of culture. In both living and nonliving forms, open thermodynamic systems allow the growth of greater levels of order so long as they have an energy gradient to feed upon. These laws of thermodynamics are likely universal all over the cosmos.

The universe also increases in complexity because it is historical. Past eras are buried in layers in the Earth. The interstellar medium is enriched with more complex atoms as stars burn and expire. A mind arriving later to the scene will have to map a more complex environment than an earlier mind, and so we should expect more processing power in later species that use the mind as a principal means of survival. Thus, without positing an élan vital or any laws other than those physics has discovered, one can talk of a universe that increases in complexity with time, so long as it has energy to power that process.

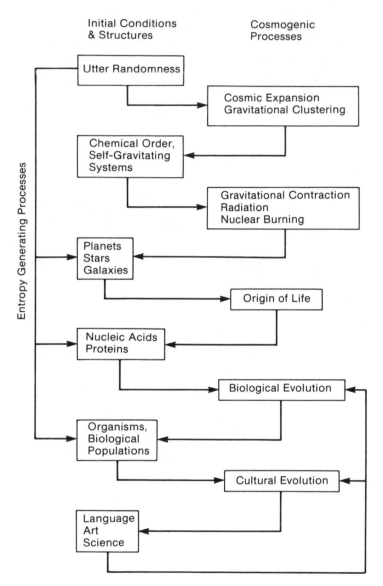

Figure 12.2 Layzer's hierarchy of time-bound order.
Each level makes the next one possible, with time running downward in the chart.
Human culture now has the ability to feed back and influence its own development
as well as biological evolution. (From David Layzer, *Cosmogenesis: The Growth of Order
in the Universe* [New York: Oxford University Press, 1990], 37. Used by permission of
the publisher and the author)

The Multiverse

How are we to understand the many improbable numbers that seem to coincide to allow a universe where life can appear? Most scientists would acknowledge the improbability of these coincidences and explain them as the result of a "multiverse." In this theory, we are only one of a multitude of universes that are continually appearing, like so many bubbles in a glass of champagne. In most of these universes, the laws and constants of nature are different, and they would produce different universes, most of them sterile. We live in a universe with observers, one that was just lucky enough to have parameters that allowed the appearance of life. The other universes would not be asking questions about origins because there would be no consciousness present to pose such questions.

Thus a kind of Darwinian logic is applied to cosmology. Just as genes vary by random mutation, leading to different structures in nature, the laws of physics vary widely in different universes. Selection favors those structures that are useful in biology, giving an illusion of design, and only those universes suitable for life will have observers to wonder at how finely tuned the laws of nature seem to be. In both cases, randomness is really the prevailing mechanism, with order and complexity as a small subset of all the possibilities.

There are several problems with applying Darwinian logic to cosmology. Complex structures in biology are the result of multiple selection events over time. No one proposes that a complex organ like the eye, for example, appears all at once by the coincidence of many favorable mutations. One can trace stages in its "complexification"—for example, from the flat patch of photoreceptors in a limpet, to an eyecup in a shelled mollusc, to a pinhole camera design found in the genus *Nautilus*, to the complex lens-type eye in squid.[30] Each structure, even the simplest one, is useful to its owner, and new functions build on those of the past. But adherents of the multiverse theory propose that all their improbabilities happen at once. There is no gradual sifting and building of one improbability on top of another. The fine-tuning of the constants of nature and the relationship of the basic forces of physics must all appear together at the birth of our universe. To encompass this immense improbability, theorists posit an infinity of universes. Surely with enough time and space the near-impossible will happen, as monkeys typing randomly at keyboards might eventually write Shakespeare.

An infinity of universes quickly generates a variety of absurdities. There are universes where there are copies of yourself; in some you finish reading this book, in others you put it down.[31] There is nothing to prevent fake universes, where a supercomputer is able to generate all the appearances of a real universe. In some universes, Hitler wins the Second World War. Can a theory that admits all possibilities really explain anything, and how can it ever be proved or disproved? Science writer Martin Gardner notes, "The stark truth is that there is not the slightest shred of reliable evidence that there is any universe other than the one we are in. No multiverse theory has so far provided a prediction that can be tested."[32]

The idea of a multiverse only shifts the problem of causation up a notch. What mechanism keeps generating bubble universes? What is the reason for this mechanism, and when did it begin? A common explanation is eternal inflation: bubble universes are continually inflating, and so too is the space between them, so by definition they will never be in contact,[33] and conveniently you will never be able to prove or disprove their existence.

There are at least nine different versions of the multiverse theory. In some, you have to posit extra dimensions in order to make the equations work. These dimensions evade detection because they are tiny and curled up, although attached to our familiar three dimensions. In one version, there are 10^{500} different forms for the extra dimensions. This is more than there have been seconds since the Big Bang (10^{18}) or the number of photons in the observable universe (10^{88}).[34]

Modern physics is in ferment, and its unwieldy complexity allows a variety of interpretations. One of the architects of the standard model in physics, Nobel laureate Steven Weinberg, admits that it "is clearly not the final answer," and writes that it "involves many features that are not dictated by fundamental principles. . . . These apparently arbitrary features include a menu of particles, a number of constants such as ratios of masses, and even the symmetries themselves. We can easily imagine that any or all of these features of the standard model might have been different."[35]

The standard model has 19 free parameters that can be adjusted to ensure agreement with experimental results.[36] These are numbers that are not predicted by the model itself, but have to be measured by experiment. For example, Newton's law of gravitation gives the mathematical form of the attraction between bodies, but the gravitational constant gives the actual strength of the force, and that number must be determined by experiment. Its magnitude is not derived from first principles, and theoretically it could vary from one universe to another.[37] This seeming arbitrariness is taken as license to posit an infinity of universes, in which the constants of nature and the forms of its laws vary. Only a few of these will achieve the happy coincidences that permit life.

CODA

In the remaining pages of this book, I would like to sketch out an alternative view from the conclusions proposed by the multiverse theory. This view borrows from some of the traditional ideas of monotheism, and will not, I realize, be to everyone's taste. I have until now confined myself to an entirely naturalistic account of events, and some will not wish to go further than that. But there comes a point where science knocks on the door of metaphysics, and in its embrace of infinities, the multiverse hypothesis calls for alternative views.

Any theory of ultimate origins eventually runs into infinities of both space and time. Where does the multiverse end, or what are the dimensions of God? The human mind has about an equal amount of trouble comprehending both these concepts. An eternity of time, as applied to the physical universe, or to a god, is equally difficult to grasp. For every explanation, the little child's question, "But why did that happen?" can be applied. Karl Popper, the noted philosopher of science, says, "There can be no explanation which is not in need of a further explanation. . . ."[38]

All things come to an end in human experience, so we have difficulty grasping eternity. Indeed, the infinite regress of physical causation suggests that an ultimate explanation may need to come at another level. Many cultures have settled on ideas of an eternal God or gods, although if you probed this idea of eternity in detail, you would probably also run into conceptual difficulties.

Religious explanations of origins are one of many possible hypotheses, and in their embrace of infinities no more intellectually objectionable, I submit, than several current theories in physics, such as the multiverse. There is, however, one great difference. The religious believer can have an emotional-spiritual relationship with the ultimate cause of the universe that a merely rationalist approach does not engender. A deeply emotional experience that goes far beyond intellectual pursuits is open to the believer. Indeed, I think very few religious people would say they believe because God provides a satisfying First Cause to the question of origins. They would speak of a deeply moving experience around which they have decided to focus their lives. Religion in that sense is more like music than analytic thinking. Science will never prove or disprove religious experience any more than it will prove or disprove the validity of music. Can science tell me why I find a Beethoven symphony deeply moving? What quantities exactly will it measure and compare? Certainly there are parts of religion that science can pass judgment on, such as the origin of humanity or the existence of a Garden of Eden. But the core experience of a loving God, like the reality of love itself, is something that I believe science will never be able to measure or pass judgment on. We might even find neurons that light up in brain images during these experiences, but it will always be possible to debate whether this comes from a kind of self-hypnosis or from a real external source.

The Anthropic Principle

The idea that our universe was deliberately tuned to allow life has sometimes been derisively called the *Anthropic Principle*. While we were one planet alone in an immense and seemingly hostile universe, it seemed like the height of egotism that all those celestial fireworks should be aimed toward us. But what of a universe that has billions upon billions of worlds, where planet making is at least as common as star formation, with habitable zones and life a strong possibility all over the cosmos? This would not be the indifferent, alien universe of existentialism, but more like a field of wildflowers springing up with the profusion and beauty that we find on Alpine meadows. Such a universe is a distinct possibility from the data presented by the latest astronomy.

A main point of this book has been the proposal that one of the peaks of the evolutionary process is a person, an emergent self, rich in communication and social relations, with an abstracting mind that is able to remake its environment. This has come to fruition in the primate lineage on our planet, but it may be part of a universal process by which nature increases in information content. If a person is one of the supreme products of the evolutionary process, it is not unreasonable to believe that a Person is the initiator of the historical process. For species on the high-information pathway, the summit of this journey will be beings who know and understand Him, as His goal is to have a relationship with them. This is an ancient view that belongs to the traditional monotheisms, but it is compatible, I submit, with the universe uncovered by modern science. One can accept entirely what modern science has to say, but still view it within this larger theistic context. In this perspective, evolution is the friend of religion and not its enemy, as many traditionalists maintain. By clinging to the letter of scripture, they are missing its spirit and how it can be reconciled with the universe revealed by modern science.

I do not mean to suggest that science can be used to prove religious beliefs. One should not expect one level of reality to prove the existence of another, and this is a problem that applies to science as well as religion. Organic chemistry does not prove there must be cells, and the existence of single-celled life does not prove there must be multicellular animals or organs like the brain or heart. Once you know a higher level of organization exists, you see that it does not violate the laws of lower levels. The heart does not violate the laws of physics and chemistry, but neither do those fields predict hearts must exist. You might extrapolate from the needs of single cells that multicellular life would require a circulatory system to provide oxygen and nutrients, but until you know the more complex level of organization, you cannot predict its existence from simpler levels. Nature cannot be used to prove the existence of God, and religion should be compatible with the findings of science.

Teilhard de Chardin

The view of nature presented here has many similarities to one proposed by paleontologist and Catholic priest Pierre Teilhard de Chardin. He saw the universe as increasing in complexity, counter to the leveling effects of entropy.[39] Increasing brain size in mammals was part of that movement, and in humans it reaches a new level of abstraction, a "power acquired by a consciousness to turn in upon itself, to take possession of itself *as of an object* . . . no longer merely to know, but to know oneself; no longer merely to know, but to know that one knows."[40] Writing in 1940, before the invention of computers and the Internet, he saw that travel and communication were bringing people together in a new way, calling for a new unity and understanding between people. A world consciousness was evolving that he called the "noosphere." This depended on not just a development of the mind, but of the heart also, of the emotions. Teilhard de Chardin looked toward a full development of the human personality because he saw the ultimate, God, as a person.

The Personalization of Nature

In chapter 8, I suggested that religion may be part of the genetic inheritance of humanity. From the time that human social life reached the tribal level, there was a need to unify people into groups larger than the family. Few animal species achieve social organizations larger than those provided by the genetic bonds between family and kin. Religion is able to achieve such groupings in humans because it creates spiritual brothers and sisters under the guidance of a spiritual father. That father may be a single God or the favored spirit in a polytheist culture. For much of history, different religions have treated each other like genetic outgroups, expressing extremes of charity to fellow members and hostility, to the point of murder, to nonbelievers.[41] Recall that at the tribal level there is constant warfare between groups and high mortality, especially among young males.[42] Any behavior that welded people together into more cohesive groups and instilled a spirit of sacrifice and unity of action could have important survival benefits. The readiness of suicide bombers to sacrifice for their group bears witness to how readily this spirit can be evoked. As humanity emerged from the Ice Ages and its hunter-gatherer way of life and began to settle down into villages, groups that could recruit more members and wield them into larger cooperative wholes for both social construction and warfare would have distinct advantages. We may all be the descendents of people who could feel the religious emotion because smaller, more fragmented groups simply disappeared.[43]

Thirty thousand years is plenty of time for a cultural habit to select among human genes. People who are descendents of dairy farmers stretching back about 9,000 years have lactose-tolerance genes for digesting milk as adults that individuals without a heritage of dairy culture do not have.[44] This is a clear example of culture favoring a certain set of genes. Religious practices date back at least as far as the painted caves in southern Europe, some 30,000 years ago, and perhaps became crucial for social organization at the dawn of Neolithic village life some 10,000 years ago.

Some theorists see social cohesion as the main explanation for religion.[45] Spirits were invented because they serve the purposes of society. But why does the utility of something prove it is an illusion? Eyes evolved to see light, and that keeps us from bumping into things, but that does not mean that the light or the objects are not real. There may indeed be a God and a certain kind of energy that people receive when they pray, and because that experience has benefits in terms of social life does not prove it is an illusion. Reductionist thought continually focuses on one aspect of an experience and claims it is the entire explanation. In a similar way, Freud saw religion as nothing more than a projection of the parent-child relationship. It takes no great insight to see that "Our father who art in heaven" has elements of that relationship, but that does not mean that is *all* that faith is.

Others point to the great variety of religious beliefs throughout history as evidence that there can be no central core of truth. But the fact that there have been many astronomies—Mayan, Ptolemaic, Arabian, and so forth—does not mean the stars are not really there. We can see the various religions of history as an attempt to conceptualize a certain level of reality and human experience, and we might come closer to a true understanding of that area of reality with time, just as we hope that our astronomies provide us with better and better approximations of what is truly out there in the heavens.

It is not my goal here to attempt such a theology,[46] but to indicate that religion and science can be compatible, both a search for truth at different levels of reality. They can run on together, like different melodies in a fugue, separate yet intermingling in places, together parts of a larger whole.

We stand at one of the axial points of history. Travel and communications are bringing us together as never before. We see the common humanity beneath the old divisions of nation, race, and creed, and we are searching for some new definition of ourselves.[47] We reach out through the Internet and the cell phones in our pockets, instantly connected to a World Wide Web of communication and information. But these ghostly electronic channels, I submit, are not the way to true relationship. A real community will be based, as it always has been, on shared values, and here religion may play a part, for it defines ultimate values. It defines what a people struggle for and provides identity. It is part of our emotional heritage, and perhaps is even in our genes.

A great challenge for religion in the future will be to overcome the out-group hostility that so often accompanies in-group solidarity. Is it possible to be secure enough in one's beliefs that one is not threatened by the beliefs of others? Presumably a greater truth should see how lesser truths are aspects of its own, the way modern astronomy can see that Ptolemaic astronomy misconceived data that it now sees more clearly. As someone suggested, there are not false religions unless you consider a child a false adult.

Some see the conflicting ideas of modern life and would go back to the old certainties. They are frightened by all the change around us and would go back to the old definitions. A variety of fundamentalisms have loosed their violence upon the world. To some this is further evidence that enlightened people should have nothing to do with religion. But the fire that cooks your dinner can also burn down your house; it does not mean you should have nothing to do with fire. That flame, I submit, is implanted in the human psyche, and the challenge of the future will be to express it passionately, but with a universality that includes kindness and tolerance to those who differ.

The view of nature presented here proposes that a person stands at the apex of nature. It is the culmination of the information-gathering process by which organisms adjust to a universe of change with behavioral flexibility. It is not the only course, for low-information strategies also work, and there can be many gradations in between. Nor am I necessarily referring to human beings, for this pathway may come to fruition on other worlds in different lineages where there has been time enough for evolution to realize its potential. When one considers the many improbabilities in the initial conditions of our universe, and how they seem peculiarly tuned to make life a possibility all over the vast stretches of the cosmos, perhaps personality was also there at the beginning, that it is both alpha and omega, and that a destiny was implanted in nature.

The universe began with the utmost uniformity, a hot plasma that melded all matter and energy into one and had a near maximum potential to fuel future activities. With time, it evolved extreme individuality in the form of high-information organisms that have much to express to each other and that modify their environment in complex ways. Perhaps the destiny of the universe is to find a resolution that preserves both individuality and difference, a unity that we know as love.

Notes

Overview

1. N. A. Reddy, "Cosmology: A Glimpse of the First Galaxies," *Nature* 469, no. 7331 (2011): 479–481.

2. "NASA Finds Earth-Size Planet Candidates in the Habitable Zone," NASA, February 2, 2011, http://www.nasa.gov/mission_pages/kepler/news/kepler_data_release .html (accessed March 18, 2011).

3. James F. Kasting, *How to Find a Habitable Planet* (Princeton: Princeton University Press, 2010), 296–297.

4. Matthew Arnold, Kenneth Allott, and Miriam Farris Allott, *The Poems of Matthew Arnold*, 2nd ed. (New York: Longman, 1979).

1. The Immune System: A Parable

1. Neil A. Campbell and Jane B. Reece, *Biology*, 6th ed. (San Francisco: Benjamin Cummings, 2002).

2. Ibid., 1154.

3. S. C. Stearns, *The Evolution of Life Histories* (Oxford: Oxford University Press, 1992); Brian Charlesworth, *Evolution in Age-Structured Populations*, Cambridge Studies in Mathematical Biology 1 (Cambridge: Cambridge University Press, 1980); M. S. Boyce, "Restitution of R-Selection and K-Selection as a Model of Density-Dependent Natural-Selection," *Annual Review of Ecology and Systematics* 15 (1984): 427–447.

4. Campbell and Reece, *Biology*, 1158.

5. James D. Mauseth, *Botany: An Introduction to Plant Biology*, 2nd ed. (Philadelphia: Saunders, 1995).

6. Ibid., 758.

7. Geerat J. Vermeij, *Evolution and Escalation: An Ecological History of Life* (Princeton: Princeton University Press, 1987).

8. C. P. Van Schaik and R. O. Deaner, "Life History and Cognitive Evolution in Primates," in *Animal Social Complexity: Intelligence, Culture, and Individualized Societies*, ed. F. B. M. de Waal and Peter L. Tyack (Cambridge, Mass.: Harvard University Press, 2003), 5–25; J. Kelley, "Life History and Cognitive Evolution in the Apes," in *The Evolution of Thought: Evolutionary Origins of Great Ape Intelligence*, ed. Anne E. Russon and David R. Begun (Cambridge: Cambridge University Press, 2004), 280–297.

9. R. G. Cutler, "Evolution of Longevity in Ungulates and Carnivores," *Gerontology* 25, no. 2 (1979): 69–86.

10. P. M. Bennett and P. H. Harvey, "Brain Size, Development and Metabolism in Birds and Mammals," *Journal of Zoology* 207 (1985): 491–509; P. H. Harvey, "Life-History Variation: Size and Mortality Patterns," in *Primate Life History and Evolution*, ed. C. Jean DeRousseau and Wenner-Gren Foundation for Anthropological Research (New York: Wiley-Liss, 1990), 81–88.

11. G. A. Sacher and E. Staffeld, "Relation of Gestation Time to Brain Weight for Placental Mammals—Implications for Theory of Vertebrate Growth," *American Naturalist* 108, no. 963 (1974): 593–615.

2. Voyages into Homeostasis

1. Monroe W. Strickberger, *Evolution*, 2nd ed., Jones and Bartlett Series in Biology (Sudbury, Mass.: Jones and Bartlett, 1996).

2. Walter B. Cannon, *The Wisdom of the Body* (New York: Norton, 1932), 301.

3. Ibid., 315.

4. James D. Mauseth, *Botany: An Introduction to Plant Biology*, 4th ed. (Sudbury, Mass.: Jones and Bartlett, 2009).

5. Neil A. Campbell and Jane B. Reece, *Biology*, 6th ed. (San Francisco: Benjamin Cummings, 2002).

6. William K. Purves, *Life, the Science of Biology*, 5th ed. (Sunderland, Mass.: Sinauer, 1998).

7. Some flowering plants, such as the grasses, have secondarily returned to wind pollination.

8. Loren C. Eiseley, *The Immense Journey* (New York: Random House, 1957).

9. Neil A. Campbell and Jane B. Reece, *Biology*, 8th ed. (San Francisco: Pearson Benjamin Cummings, 2009).

10. Ibid., 520.

11. M. J. Benton, *Vertebrate Palaeontology*, 3rd ed. (Malden, Mass.: Blackwell Science, 2005).

12. D. Sol et al., "Brain Size Predicts the Success of Mammal Species Introduced into Novel Environments," *American Naturalist* 172, suppl 1 (2008): S63–S71.

13. D. Sol et al., "Big-Brained Birds Survive Better in Nature." *Proceedings of the Royal Society B—Biological Sciences* 274, no. 1611 (2007): 763–769.

14. D. Sol, L. Lefebvre, and J. D. Rodriguez-Teijeiro, "Brain Size, Innovative Propensity and Migratory Behaviour in Temperate Palaearctic Birds," *Proceedings of the Royal Society B—Biological Sciences* 272, no. 1571 (2005): 1433–1441.

15. D. Sol, "Revisiting the Cognitive Buffer Hypothesis for the Evolution of Large Brains," *Biology Letters* 5, no. 1 (2009): 130–133.

16. Ibid., 130.

17. Harry J. Jerison, *Evolution of the Brain and Intelligence* (New York: Academic, 1973).

18. O. R. Bininda-Emonds et al., "The Delayed Rise of Present-Day Mammals," *Nature* 446, no. 7135 (2007): 507–512; J. Alroy, P. L. Koch, and J. C. Zachos, "Global Climate Change and North American Mammalian Evolution," *Paleobiology* 26, no. 4 (2000): 259–288.

19. J. W. Brown, R. B. Payne, and D. P. Mindell, "Nuclear DNA Does Not Reconcile 'Rocks' and 'Clocks' in Neoaves: A Comment on Ericson et al.," *Biology Letters* 3, no. 3 (2007): 257–259; discussion 60–61.

20. Benton, *Vertebrate Palaeontology*, 286.

21. Ibid., 251.

22. Georg F. Striedter, *Principles of Brain Evolution* (Sunderland, Mass.: Sinauer, 2005); Z. X. Luo, A. W. Crompton, and A. L. Sun, "A New Mammaliaform from the Early Jurassic and Evolution of Mammalian Characteristics," *Science* 292, no. 5521 (2001): 1535–1540.

23. Zofia Kielan-Jaworowska, Richard Cifelli, and Zhe-Xi Luo, *Mammals from the Age of Dinosaurs: Origins, Evolution, and Structure* (New York: Columbia University Press, 2004).

24. J. H. Kaas, "Reconstructing the Organization of Neocortex of the First Mammals and Subsequent Modifications," in *Evolution of Nervous Systems: A Comprehensive Reference*, vol. 3, *Mammals*, ed. Jon H. Kaas and Theodore H. Bullock (Amsterdam: Elsevier Academic, 2007), 27–48.

25. Striedter, *Principles of Brain Evolution*, 107.

26. R. Potts, "Paleoenvironmental Basis of Cognitive Evolution in Great Apes," *American Journal of Primatology* 62, no. 3 (2004): 209–228.

27. Ibid., 223.

28. R. Potts, "Variability Selection in Hominid Evolution," *Evolutionary Anthropology* 7, no. 3 (1998): 81–96.

29. Ibid., 93.

30. Potts, "Paleoenvironmental Basis of Cognitive Evolution," 225.

31. J. C. McElwain and S. W. Punyasena, "Mass Extinction Events and the Plant Fossil Record," *Trends in Ecology and Evolution* 22, no. 10 (2007): 548–557.

32. David E. Sadava, *Life, the Science of Biology*, 8th ed. (Sunderland, Mass.: Sinauer; Gordonsville, Va.: Freeman, 2008); Campbell and Reece, *Biology*, 8th ed., 524.

33. W. L. Crepet and K. J. Niklas, "Darwin's Second 'Abominable Mystery': Why Are There so Many Angiosperm Species?" *American Journal of Botany* 96, no. 1 (2009): 366–381.

34. Ibid., 372.

35. Campbell and Reece, *Biology*, 8th ed., 794.

36. F. Berendse and M. Scheffer, "The Angiosperm Radiation Revisited, an Ecological Explanation for Darwin's 'Abominable Mystery'," *Ecology Letters* 12, no. 9 (2009): 865–872.

37. J. B. Heo and S. Sung. "Vernalization-Mediated Epigenetic Silencing by a Long Intronic Noncoding RNA," *Science* 331, no. 6013 (2011): 76–79.

38. E. Kejnovsky, I. J. Leitch, and A. R. Leitch, "Contrasting Evolutionary Dynamics Between Angiosperm and Mammalian Genomes," *Trends in Ecology and Evolution* 24, no. 10 (2009): 572–582.

39. D. E. Soltis et al., "Polyploidy and Angiosperm Diversification," *American Journal of Botany* 96, no. 1 (2009): 336–348.

40. Berendse and Scheffer, "Angiosperm Radiation," 866.

41. Ibid., 870.

42. W. K. Cornwell et al., "Plant Species Traits Are the Predominant Control on Litter Decomposition Rates Within Biomes Worldwide," *Ecology Letters* 11, no. 10 (2008): 1065–1071.

43. Benton, *Vertebrate Palaeontology*, 190.

44. Peter H. Raven, Ray Franklin Evert, and Susan E. Eichhorn, *Biology of Plants*, 7th ed. (New York: Freeman, 2005).

45. J. Zalasiewicz and M. Williams, "A Geological History of Climate Change," in *Climate Change: Observed Impacts on Planet Earth*, ed. T. M. Letcher (Amsterdam: Elsevier, 2009), 127–142; Peter W. Skelton, ed., *The Cretaceous World* (Cambridge: Open University, Cambridge University Press, 2003).

46. Purves, *Life*, 434–435.

47. N. Lane and W. Martin, "The Energetics of Genome Complexity," *Nature* 467, no. 7318 (2010): 929–934.

48. Ibid., 929–933.

49. D. E. Canfield, S. W. Poulton, and G. M. Narbonne, "Late-Neoproterozoic Deep-Ocean Oxygenation and the Rise of Animal Life," *Science* 315, no. 5808 (2007): 92–95.

50. R. A. Berner, J. M. Vandenbrooks, and P. D. Ward, "Evolution: Oxygen and Evolution," *Science* 316, no. 5824 (2007): 557–558.

51. David E. Sadava et al., *Life, the Science of Biology*, 9th ed. (Sunderland, Mass.: Sinauer, 2011), 523–532.

52. P. G. Falkowski et al., "The Rise of Oxygen over the Past 205 Million Years and the Evolution of Large Placental Mammals," *Science* 309, no. 5744 (2005): 2202–2204.

53. Striedter, *Principles of Brain Evolution*, 259.

54. R. I. Dunbar and S. Shultz, "Understanding Primate Brain Evolution," *Philosophical Transactions of the Royal Society B—Biological Sciences* 362, no. 1480 (2007): 649–658.

55. K. Isler and C. P. van Schaik, "The Expensive Brain: A Framework for Explaining Evolutionary Changes in Brain Size," *Journal of Human Evolution* (2009): 392–400.

56. M. J. Benton, "Diversification and Extinction in the History of Life," *Science* 268, no. 5207 (1995): 52–58.

3. Information Content

1. Michael Lynch, *The Origins of Genome Architecture* (Sunderland, Mass.: Sinauer, 2007).

2. N. Lane and W. Martin, "The Energetics of Genome Complexity," *Nature* 467, no. 7318 (2010): 929–934.

3. L. K. Fritz-Laylin et al., "The Genome of Naegleria Gruberi Illuminates Early Eukaryotic Versatility," *Cell* 140, no. 5 (2010): 631–642.

4. Lane and Martin, "Energetics of Genome Complexity," 931.

5. N. Molina and E. van Nimwegen, "Scaling Laws in Functional Genome Content Across Prokaryotic Clades and Lifestyles," *Trends in Genetics* 25, no. 6 (2009): 243–247.

6. E. van Nimwegen, "Scaling Laws in the Functional Content of Genomes," *Trends in Genetics* 19, no. 9 (2003): 479–484.

7. Mihaela Pertea and Steven L. Salzberg, "Between a Chicken and a Grape: Estimating the Number of Human Genes," *Genome Biology* 11 (2010): 206.

8. Ensembl project genome database, http://uswest.ensembl.org/index.html (accessed May 23, 2011).

9. K. Nowick and L. Stubbs, "Lineage-Specific Transcription Factors and the Evolution of Gene Regulatory Networks," *Briefings in Functional Genomics* 9, no. 1 (2010): 65–78.

10. C. Vogel and C. Chothia, "Protein Family Expansions and Biological Complexity," *PLoS Computational Biology* 2, no. 5 (2006): 370–382.

11. Ibid., 379.

12. Lynch, *Origins of Genome Architecture*, 52.

13. N. J. Martinez and A. J. M. Walhout, "The Interplay Between Transcription Factors and Micrornas in Genome-Scale Regulatory Networks," *Bioessays* 31, no. 4 (2009): 435–445.

14. C. R. Marshall and J. W. Valentine, "The Importance of Preadapted Genomes in the Origin of the Animal Bodyplans and the Cambrian Explosion," *Evolution* 64, no. 5 (2010): 1189–1201.

15. S. S. Merchant et al., "The Chlamydomonas Genome Reveals the Evolution of Key Animal and Plant Functions," *Science* 318, no. 5848 (2007): 245–250.

16. S. E. Prochnik et al., "Genomic Analysis of Organismal Complexity in the Multicellular Green Alga Volvox Carteri," *Science* 329, no. 5988 (2010): 223–226.

17. N. King et al., "The Genome of the Choanoflagellate Monosiga Brevicollis and the Origin of Metazoans," *Nature* 451, no. 7180 (2008): 783–788.

18. Ibid., 785–787.

19. M. Srivastava et al., "The Amphimedon Queenslandica Genome and the Evolution of Animal Complexity," *Nature* 466, no. 7307 (2010): 720–726.

20. Ibid., 724.

21. J. A. Chapman et al., "The Dynamic Genome of Hydra," *Nature* 464, no. 7288 (2010): 592–596.

22. Ibid., 595.

23. C. Larroux et al., "Genesis and Expansion of Metazoan Transcription Factor Gene Classes," *Molecular Biology and Evolution* 25, no. 5 (2008): 980–996.

24. L. Z. Holland et al., "The Amphioxus Genome Illuminates Vertebrate Origins and Cephalochordate Biology," *Genome Research* 18, no. 7 (2008): 1100–1111.

25. N. H. Putnam et al., "The Amphioxus Genome and the Evolution of the Chordate Karyotype," *Nature* 453, no. 7198 (2008): 1064–1071.

26. Ibid., 1069.

27. E. H. Davidson and D. H. Erwin, "An Integrated View of Precambrian Eumetazoan Evolution," *Cold Spring Harbor Symposia on Quantitative Biology* 74 (2009): 65–80.

28. T. Vavouri and B. Lehner, "Conserved Noncoding Elements and the Evolution of Animal Body Plans," *Bioessays* 31, no. 7 (2009): 727–735.

29. C. R. Marshall, "Explaining the Cambrian 'Explosion' of Animals," *Annual Review of Earth and Planetary Sciences* 34 (2006): 355–384.

30. J. Zeitlinger and A. Stark, "Developmental Gene Regulation in the Era of Genomics," *Developmental Biology* 339, no. 2 (2010): 232–233.

31. Davidson and Erwin, "Precambrian Eumetazoan Evolution," 10.

32. S. B. Carroll, "Evo-Devo and an Expanding Evolutionary Synthesis: A Genetic Theory of Morphological Evolution," *Cell* 134, no. 1 (2008): 25–36.

33. S. B. Carroll, "Chance and Necessity: The Evolution of Morphological Complexity and Diversity," *Nature* 409, no. 6823 (2001): 1102–1109.

34. Ibid., 1108.

35. J. D. Richter, "Think You Know How Mirnas Work? Think Again," *Nature Structural and Molecular Biology* 15, no. 4 (2008): 334–336.

36. miRBase database, http://www.mirbase.org/cgi-bin/browse.pl (accessed May 23, 2011).

37. K. J. Peterson, M. R. Dietrich, and M. A. McPeek, "MicroRNAs and Metazoan Macroevolution: Insights into Canalization, Complexity, and the Cambrian Explosion," *Bioessays* 31, no. 7 (2009): 736–747.

38. A. M. Heimberg et al., "MicroRNAs and the Advent of Vertebrate Morphological Complexity," *Proceedings of the National Acadacemy of Sciences of the United States of America* 105, no. 8 (2008): 2946–2950.

39. B. M. Wheeler et al., "The Deep Evolution of Metazoan Micrornas," *Evolution and Development* 11, no. 1 (2009): 50–68.

40. Peterson, Dietrich, and McPeek, "MicroRNAs and Metazoan Macroevolution," 742.

41. Zeitlinger and Stark, "Developmental Gene Regulation," 230–239.

42. Ibid., 231.

43. F. A. Azevedo et al., "Equal Numbers of Neuronal and Nonneuronal Cells Make the Human Brain an Isometrically Scaled-up Primate Brain," *Journal of Comparative Neurology* 513, no. 5 (2009): 532–541.

44. Georg F. Striedter, *Principles of Brain Evolution* (Sunderland, Mass.: Sinauer, 2005); V. S. Ramachandran, *Encyclopedia of the Human Brain* (San Diego: Academic, 2002).

45. Bruce Alberts, John H. Wilson, and Tim Hunt, *Molecular Biology of the Cell*, 5th ed. (New York: Garland Science, 2008).

46. Carl Sagan, *The Dragons of Eden: Speculations on the Evolution of Human Intelligence* (New York: Random House, 1977).

47. Howard K. Bloom, *The Global Brain: The Evolution of Mass Mind from the Big Bang to the 21st Century* (New York: Wiley, 2000).

48. *World Wide Web*, s.v. "Statistics," http://en.wikipedia.org/wiki/World_Wide_Web#Statistics (accessed May 23, 2011).

49. Joseph E. LeDoux, *Synaptic Self: How Our Brains Become Who We Are* (New York: Viking, 2002).

50. M. H. Johnson, "Cortical Mechanisms of Cognitive Development," in *Cognitive Neuroscience: A Reader*, ed. Michael S. Gazzaniga (Malden, Mass.: Blackwell, 2000), 241–260.

51. LeDoux, *Synaptic Self*, 77.

52. J. Bourgeois, P. S. Goldman-Rakic, and P. Rakic, "Formation, Elimination, and Stabilization of Synapses in the Primate Cerebral Cortex," in *The New Cognitive Neurosciences*, ed. Michael S. Gazzaniga (Cambridge, Mass.: MIT Press, 2000), 45–54.

53. Harry J. Jerison, *Evolution of the Brain and Intelligence* (New York: Academic, 1973).

54. H. J. Jerison, "Epilogue. The Study of Primate Brain Evolution: Where Do We Go from Here?" in *Evolutionary Anatomy of the Primate Cerebral Cortex*, ed. Dean Falk and Kathleen Rita Gibson (Cambridge: Cambridge University Press, 2001), 305–336.

55. Striedter, *Principles of Brain Evolution*, 93.

56. John Frederick Eisenberg, *The Mammalian Radiations: An Analysis of Trends in Evolution, Adaptation, and Behavior* (Chicago: University of Chicago Press, 1981). The general form of the regression line is given by $y = ax^b$, where y is brain weight and x represents body weight. When logs are taken of both sides, b becomes the slope and a is the y intercept.

57. Striedter, *Principles of Brain Evolution*, 96–98.

58. Eisenberg, *Mammalian Radiations*, 498–502.

59. Jerison, *Evolution of the Brain and Intelligence*, 395.

60. S. Herculano-Houzel, "Brains Matter, Bodies Maybe Not: The Case for Examining Neuron Numbers Irrespective of Body Size," *Annals of the New York Academy of Sciences* 1225, no. 1 (2011): 191–199.

61. H. J. Jerison, "The Evolution of Neural and Behavioral Complexity," in *Brain Evolution and Cognition*, ed. Gerhard Roth and Mario F. Wullimann (New York: Wiley, 2001), 523–553.

62. Ibid., 539.

63. Leonard B. Radinsky, *The Evolution of Vertebrate Design* (Chicago: University of Chicago Press, 1987).

64. Jerison, "Evolution of Neural and Behavioral Complexity," 539.

65. A. Portmann and W. Stingelin, "The Central Nervous System," in *Biology and Comparative Physiology of Birds*, ed. Alan John Marshall (New York: Academic, 1961), 1–36.

66. S. Herculano-Houzel, "Coordinated Scaling of Cortical and Cerebellar Numbers of Neurons," *Frontiers in Neuroanatomy* 4 (2010): 1–8.

67. A. Schuz, "What Can the Cerebral Cortex Do Better Than Other Parts of the Brain?" in *Brain Evolution and Cognition*, ed. Roth and Wullimann, 491–500.

68. M. A. Hofman, "Evolution and Complexity of the Human Brain: Some Organizing Principles," in *Brain Evolution and Cognition*, ed. Roth and Wullimann, 501–521.

69. P. Rakic and D. R. Kornack, "Neocortical Expansion and Elaboration During Primate Evolution: A View from Neroembryology," in *Evolutionary Anatomy of the Primate Cerebral Cortex*, ed. Falk and Gibson, 30–56.

70. P. Rakic, "Setting the Stage for Cognition: Genesis of the Primate Cerebral Cortex," in *New Cognitive Neurosciences*, ed. Gazzaniga, 7–22.

71. Scott F. Gilbert, *Developmental Biology*, 7th ed. (Sunderland, Mass.: Sinauer, 2003).

72. S. Herculano-Houzel, "The Human Brain in Numbers: A Linearly Scaled-up Primate Brain," *Frontiers in Human Neuroscience* 3 (2009): 31.

73. Chimpanzee Sequencing and Analysis Consortium, "Initial Sequence of the Chimpanzee Genome and Comparison with the Human Genome," *Nature* 437, no. 7055 (2005): 69–87.

74. S. Dorus et al., "Accelerated Evolution of Nervous System Genes in the Origin of Homo Sapiens," *Cell* 119, no. 7 (2004): 1027–1040.

75. W. Enard et al., "Intra- and Interspecific Variation in Primate Gene Expression Patterns," *Science* 296, no. 5566 (2002): 340–343.

76. Fritz-Laylin et al., "Genome of Naegleria Gruberi," 638.

77. T. W. Nilsen and B. R. Graveley, "Expansion of the Eukaryotic Proteome by Alternative Splicing," *Nature* 463, no. 7280 (2010): 457–463.

78. T. Kawashima et al., "Domain Shuffling and the Evolution of Vertebrates," *Genome Research* 19, no. 8 (2009): 1393–1403.

79. Douglas J. Futuyma, *Evolution*, 2nd ed. (Sunderland, Mass.: Sinauer, 2009).

4. What Is a Big Brain Good For?

1. Sara J. Shettleworth, *Cognition, Evolution, and Behavior* (New York: Oxford University Press, 1998); E. M. Macphail, *Brain and Intelligence in Vertebrates*, Oxford Science Publications (New York: Clarendon, 1982).

2. R. I. M. Dunbar, "The Social Brain Hypothesis," *Evolutionary Anthropology* 6, no. 5 (1998): 178–190.

3. F. B. M. de Waal, *Chimpanzee Politics: Power and Sex Among Apes* (Baltimore: Johns Hopkins University Press, 1989).

4. F. B. M. de Waal and F. Aureli, "Consolation, Reconciliation, and a Possible Cognitive Difference Between Macaques and Chimpanzees," in *Reaching into Thought: The Minds of the Great Apes*, ed. Anne E. Russon, Kim A. Bard, and Sue Taylor Parker (Cambridge: Cambridge University Press, 1996), 81–110.

5. Dunbar, "Social Brain Hypothesis," 185–186.

6. F. J. Perez-Barberia, S. Shultz, and R. I. Dunbar, "Evidence for Coevolution of Sociality and Relative Brain Size in Three Orders of Mammals," *Evolution* 61, no. 12 (2007): 2811–2821.

7. R. I. Dunbar and S. Shultz, "Evolution in the Social Brain," *Science* 317, no. 5843 (2007): 1344–1347.

8. N. J. Emery et al., "Cognitive Adaptations of Social Bonding in Birds," *Philosophical Transactions of the Royal Society* of London—*Biological Sciences* 362, no. 1480 (2007): 489–505.

9. Richard W. Byrne and Andrew Whiten, eds., *Machiavellian Intelligence: Social Expertise and the Evolution of Intellect in Monkeys, Apes, and Humans*, Oxford Science Publications (Oxford: Clarendon, 1988); Andrew Whiten and Richard W. Byrne, eds., *Machiavellian Intelligence II: Extensions and Evaluations* (Cambridge: Cambridge University Press, 1997).

10. R. W. Byrne and N. Corp, "Neocortex Size Predicts Deception Rate in Primates," *Proceedings of the Royal Society of London B—Biological Sciences* 271, no. 1549 (2004): 1693–1699.

11. R. A. Barton, "Primate Brain Evolution: Cognitive Demands of Foraging or of Social Life?" in *On the Move: How and Why Animals Travel in Groups*, ed. Sue Boinski and Paul Alan Garber (Chicago: University of Chicago Press, 2000), 204–237.

12. Marc D. Hauser, *The Evolution of Communication* (Cambridge, Mass.: MIT Press, 1996).

13. Ibid., 249.

14. P. H. Harvey, R. D. Martin, and T. Clutton-Brock, "Life Histories in Comparative Perspective," in *Primate Societies*, ed. Barbara B. Smuts (Chicago: University of Chicago Press, 1987), 181–196.

15. P. H. Harvey, T. H. Clutton-Brock, and G. M. Mace, "Brain Size and Ecology in Small Mammals and Primates," *Proceedings of the National Academy of Sciences of the United States of America* 77, no. 7 (1980): 4387–4389.

16. John Frederick Eisenberg, *The Mammalian Radiations: An Analysis of Trends in Evolution, Adaptation, and Behavior* (Chicago: University of Chicago Press, 1981).

17. K. Milton, "Foraging Behaviour and the Evolution of Primate Intelligence," in *Machiavellian Intelligence*, ed. Byrne and Whiten, 285–306; K. Milton, "Quo Vadis? Tactics of Food Search and Group Movement in Primates and Other Animals," in *On the Move*, ed. Boinski and Garber, 375–417.

18. Milton, "Quo Vadis?" 414–415.

19. L. Lefebvre, S. M. Reader, and D. Sol, "Brains, Innovations and Evolution in Birds and Primates," *Brain, Behavior and Evolution* 63, no. 4 (2004): 233–246.

20. Dunbar and Shultz, "Evolution in the Social Brain," 1345.

21. A. N. Iwaniuk, L. Lefebvre, and D. R. Wylie, "The Comparative Approach and Brain-Behaviour Relationships: A Tool for Understanding Tool Use," *Canadian Journal of Experimental Psychology* 63, no. 2 (2009): 150–159.

22. L. Lefebvre, N. Nicolakakis, and D. Boire, "Tools and Brains in Birds," *Behaviour* 139 (2002): 939–973.

23. D. Sol and L. Lefebvre, "Behavioural Flexibility Predicts Invasion Success in Birds Introduced to New Zealand," *Oikos* 90, no. 3 (2000): 599–605.

24. D. Sol et al., "Big Brains, Enhanced Cognition, and Response of Birds to Novel Environments," *Proceedings of the National Academy of Sciences of the United States of America* 102, no. 15 (2005): 5460–5465.

25. Lefebvre, Reader, and Sol, "Brains, Innovations and Evolution," 241–242.

26. D. Sol, L. Lefebvre, and J. D. Rodriguez-Teijeiro, "Brain Size, Innovative Propensity and Migratory Behaviour in Temperate Palaearctic Birds," *Proceedings of the Royal Society B—Biological Sciences* 272, no. 1571 (2005): 1433–1441.

27. S. Shultz et al., "Brain Size and Resource Specialization Predict Long-Term Population Trends in British Birds," *Proceedings of the Royal Society B—Biological Sciences* 272, no. 1578 (2005): 2305–2311.

28. C. Schuck-Paim, W. J. Alonso, and E. B. Ottoni, "Cognition in an Ever-Changing World: Climatic Variability Is Associated with Brain Size in Neotropical Parrots," *Brain, Behavior and Evolution* 71, no. 3 (2008): 200–215.

29. A. A. Maklakov et al., "Brains and the City: Big-Brained Passerine Birds Succeed in Urban Environments," *Biology Letters* (2011), in press. doi:10.1098.

30. A. C. Milner and S. A. Walsh, "Avian Brain Evolution: New Data from Palaeogene Birds (Lower Eocene) from England," *Zoological Journal of the Linnean Society* 155, no. 1 (2009): 198–219.

31. J. S. Wyles, J. G. Kunkel, and A. C. Wilson, "Birds, Behavior, and Anatomical Evolution," *Proceedings of the National Academy of Sciences of the United States of America* 80, no. 14 (1983): 4394–4397.

32. L. M. Cherry et al., "Body Shape Metrics and Organismal Evolution," *Evolution* 36, no. 5 (1982): 914–933.

33. Alister Clavering Hardy, *The Living Stream: A Restatement of Evolution Theory and Its Relation to the Spirit of Man* (London: Collins, 1965); R. L. Neubauer, "Super-Normal Length Song Preferences of Female Zebra Finches (Taeniopygia Guttata) and a Theory of the Evolution of Bird Song," *Evolutionary Ecology* 13, no. 4 (1999): 365–380; Eytan Avital and Eva Jablonka, *Animal Traditions: Behavioural Inheritance in Evolution* (Cambridge: Cambridge University Press, 2000).

34. H. J. Jerison, "What Fossils Tell Us About the Evolution of the Neocortex," in *Evolution of Nervous Systems: A Comprehensive Reference*, vol. 3, *Mammals*, ed. Jon H. Kaas and Theodore H. Bullock (Amsterdam: Elsevier Academic, 2007), 1–13.

35. P. G. Falkowski et al., "The Rise of Oxygen over the Past 205 Million Years and the Evolution of Large Placental Mammals," *Science* 309, no. 5744 (2005): 2202–2204.

36. D. M. Rumbaugh, "Competence, Cortex, and Primate Models: A Comparative Primate Perspective," in *Development of the Prefrontal Cortex: Evolution, Neurobiology, and Behavior*, ed. Norman A. Krasnegor, G. Reid Lyon, and Patricia S. Goldman-Rakic (Baltimore: Brookes, 1997), 117–139.

37. B. Wilson, N. J. Mackintosh, and R. A. Boakes, "Transfer of Relational Rules in Matching and Oddity Learning by Pigeons and Corvids," *Quarterly Journal of Experimental Psychology Section B—Comparative and Physiological Psychology* 37, no. 4 (1985): 313–332.

38. Shettleworth, *Cognition, Evolution, and Behavior*, 212.

39. Macphail, *Brain and Intelligence in Vertebrates*, 165–167.

40. A. N. Iwaniuk, J. E. Nelson, and S. M. Pellis, "Do Big-Brained Animals Play More? Comparative Analyses of Play and Relative Brain Size in Mammals," *Journal of Comparative Psychology* 115, no. 1 (2001): 29–41.

41. Thomas George Power, *Play and Exploration in Children and Animals* (Mahwah, N.J.: Erlbaum, 2000).

42. G. M. Burghardt, "Play," in *Comparative Psychology: A Handbook*, ed. Gary Greenberg and Maury M. Haraway (New York: Garland, 1998), 725–735.

43. Ibid., 137.

44. J. C. Ortega and M. Bekoff, "Avian Play—Comparative Evolutionary and Developmental Trends," *The Auk* 104, no. 2 (1987): 338–341.

45. T. M. Caro, "Short-Term Costs and Correlates of Play in Cheetahs," *Animal Behaviour* 49, no. 2 (1995): 333–345.

46. Iwaniuk, Nelson, and Pellis, "Do Big-Brained Animals Play More?" 37.

47. Power, *Play and Exploration in Children and Animals*, 195.

48. Ibid., 43.

49. J. J. Negro et al., "Captive Fledgling American Kestrels Prefer to Play with Objects Resembling Natural Prey," *Animal Behaviour* 52 (1996): 707–714.

50. Gordon M. Burghardt, *The Genesis of Animal Play: Testing the Limits* (Cambridge, Mass.: MIT Press, 2005).

51. Power, *Play and Exploration in Children and Animals*, 45.

52. Caro, "Short-Term Costs and Correlates of Play in Cheetahs," 339.

53. M. Biben, "Squirrel Monkey Playfighting: Making the Case for a Cognitive Training Function for Play," in *Animal Play: Evolutionary, Comparative, and Ecological Prespectives*, ed. Marc Bekoff and John A. Byers (Cambridge: Cambridge University Press, 1998), 161–182.

54. S. M. Pellis and V. C. Pellis, "Structure-Function Interface in the Analysis of Play Fighting," in *Animal Play*, ed. Bekoff and Byers, 115–140.

55. M. Bekoff, "The Evolution of Animal Play, Emotions, and Social Morality: On Science, Theology, Spirituality, Personhood, and Love," *Zygon* 36 (2001): 615–655.

56. Sue Taylor Parker and Michael L. McKinney, *Origins of Intelligence: The Evolution of Cognitive Development in Monkeys, Apes, and Humans* (Baltimore: Johns Hopkins University Press, 1999).

57. T. H. Joffe, "Social Pressures Have Selected for an Extended Juvenile Period in Primates," *Journal of Human Evolution* 32, no. 6 (1997): 593–605.

58. N. L. Barrickman et al., "Life History Costs and Benefits of Encephalization: A Comparative Test Using Data from Long-Term Studies of Primates in the Wild," *Journal of Human Evolution* 54, no. 5 (2008): 568–590; K. Isler and C. P. van Schaik, "The Expensive Brain: A Framework for Explaining Evolutionary Changes in Brain Size," *Journal of Human Evolution* 57, no. 4 (2009): 392–400; B. E. Saether, "Pattern of Covariation Between Life-History Traits of European Birds," *Nature* 331, no. 6157 (1988): 616–617.

59. C. Gonzalez-Lagos, D. Sol, and S. M. Reader, "Large-Brained Mammals Live Longer," *Journal of Evolutionary Biology* 23, no. 5 (2010): 1064–1074.

60. D. Sol, "Revisiting the Cognitive Buffer Hypothesis for the Evolution of Large Brains," *Biology Letters* 5, no. 1 (2009): 130–133.

61. Parker and McKinney, *Origins of Intelligence*, 251–252.

62. E. R. Sowell et al., "In Vivo Evidence for Post-Adolescent Brain Maturation in Frontal and Striatal Regions," *Nature Neuroscience* 2, no. 10 (1999): 859–861.

63. Parker and McKinney, *Origins of Intelligence*, 223.

64. L. Marino, "Big Brains Do Matter in New Environments," *Proceedings of the National Academy of Sciences of the United States of America* 102, no. 15 (2005): 5306–5307.

5. A Constellation of Qualities

1. E. A. Brenowitz, "Birdsong: Integrating Physics, Physiology, and Behavior," *Journal of Comparative Physiology A—Neuroethology Sensory Neural and Behavioral Physiology* 188, no. 11-2 (2002): 827–828; S. Iyengar and S. W. Bottjer, "The Role of Auditory Experience in the Formation of Neural Circuits Underlying Vocal Learning in Zebra Finches," *Journal of Neuroscience* 22, no. 3 (2002): 946–958.

2. T. Matsuzawa, "Chimpanzee Intelligence in Nature and in Captivity: Isomorphism of Symbol Use and Tool Use," in *Great Ape Societies*, ed. W. C. McGrew, L. F. Marchant, and T. Nishida (Cambridge: Cambridge University Press, 1996), 196–209.

3. C. Fajardo et al., "Von Economo Neurons Are Present in the Dorsolateral (Dysgranular) Prefrontal Cortex of Humans," *Neuroscience Letters* 435, no. 3 (2008): 215–218.

4. J. Allman and A. Hasenstaub, "Brains, Maturation Times, and Parenting," *Neurobiology of Aging* 20, no. 4 (1999): 447–454.

5. Jane Goodall, *The Chimpanzees of Gombe: Patterns of Behavior* (Cambridge, Mass.: Belknap, 1986).

6. Christophe Boesch and Hedwige Boesch-Achermann, *The Chimpanzees of the Taï Forest: Behavioural Ecology and Evolution* (Oxford: Oxford University Press, 2000).

7. Frans de Waal, *Good Natured: The Origins of Right and Wrong in Humans and Other Animals* (Cambridge, Mass.: Harvard University Press, 1996).

8. Ibid., 129.

9. Boesch and Boesch-Achermann, *Chimpanzees of the Taï Forest*, 239.

10. Michael Tomasello and Josep Call, *Primate Cognition* (New York: Oxford University Press, 1997).

11. Goodall, *Chimpanzees of Gombe*, 127.

12. Vernon Reynolds, *The Chimpanzees of the Budongo Forest: Ecology, Behaviour, and Conservation*, Oxford Biology (Oxford: Oxford University Press, 2005).

13. Tomasello and Call, *Primate Cognition*, 281.

14. Ibid., 246.

15. Duane M. Rumbaugh and David A. Washburn, *Intelligence of Apes and Other Rational Beings*, Current Perspectives in Psychology (New Haven: Yale University Press, 2003).

16. Ibid., 109.

17. Tomasello and Call, *Primate Cognition*, 266–267.

18. Rumbaugh and Washburn, *Intelligence of Apes*, 141–142.

19. Andrew Whiten and Richard W. Byrne, eds., *Machiavellian Intelligence II: Extensions and Evaluations* (Cambridge: Cambridge University Press, 1997).

20. Goodall, *Chimpanzees of Gombe*, 577.

21. de Waal, *Good Natured*, 44.

22. B. Hare, J. Call, and M. Tomasello, "Chimpanzees Deceive a Human Competitor by Hiding," *Cognition* 101, no. 3 (2006): 495–514.

23. M. Tomasello, J. Call, and B. Hare, "Chimpanzees Understand Psychological States—The Question Is Which Ones and to What Extent," *Trends in Cognitive Sciences* 7, no. 4 (2003): 153–156.

24. A. P. Melis, J. Call, and M. Tomasello, "Chimpanzees (*Pan troglodytes*) Conceal Visual and Auditory Information from Others," *Journal of Comparative Psychology* 120, no. 2 (2006): 154–162.

25. Boesch and Boesch-Achermann, *Chimpanzees of the Taï Forest*, 193.

26. Ibid., 202–206.

27. Matsuzawa, "Chimpanzee Intelligence," 203–205.

28. Tetsuro Matsuzawa, "Primate Foundations of Human Intelligence: A View of Tool Use in Nonhuman Primates and Fossil Hominids," in *Primate Origins of Human Cognition and Behavior*, ed. Tetsuro Matsuzawa (Tokyo: Springer, 2001), 3–25.

29. S. M. Brewer and W. C. Mcgrew, "Chimpanzee Use of a Tool-Set to Get Honey," *Folia Primatologica* 54, no. 1-2 (1990): 100–104.

30. C. Sanz, D. Morgan, and S. Gulick, "New Insights into Chimpanzees, Tools, and Termites from the Congo Basin," *American Naturalist* 164, no. 5 (2004): 567–581.

31. Ibid., 575.

32. Matsuzawa, "Primate Foundations," 15.

33. Matsuzawa, "Chimpanzee Intelligence," 204.

34. K. E. Langergraber et al., "Genetic and 'Cultural' Similarity in Wild Chimpanzees," *Proceedings of the Royal Society B—Biological Sciences* 278, no. 1704 (2011): 408–416.

35. Boesch and Boesch-Achermann, *Chimpanzees of the Taï Forest*, 197.

36. Goodall, *Chimpanzees of Gombe*, 426.

37. T. Nishida, "Individuality and Flexibility of Cultural Patterns in Chimpanzees," in *Animal Social Complexity: Intelligence, Culture, and Individualized Societies*, ed. F. B. M. de Waal and Peter L. Tyack (Cambridge, Mass.: Harvard University Press, 2003), 392–413.

38. D. Biro et al., "Cultural Innovation and Transmission of Tool Use in Wild Chimpanzees: Evidence from Field Experiments," *Animal Cognition* 6, no. 4 (2003): 213–223.

39. A. Whiten et al., "Cultures in Chimpanzees," *Nature* 399, no. 6737 (1999): 682–685.

40. Boesch and Boesch-Achermann, *Chimpanzees of the Taï Forest*, 262.

41. George Page, *Inside the Animal Mind* (New York: Doubleday, 1999).

42. Boesch and Boesch-Achermann, *Chimpanzees of the Taï Forest*, 113–116.

43. Goodall, *Chimpanzees of Gombe*, 366.

44. Ibid., 363.

45. Ibid., 361–364.

46. de Waal, *Good Natured*, 43.

47. Boesch and Boesch-Achermann, *Chimpanzees of the Taï Forest*, 247.

48. Ibid., 250.

49. F. Warneken et al., "Spontaneous Altruism by Chimpanzees and Young Children," *PLoS Biology* 5, no. 7 (2007): e184.

50. K. E. Langergraber, J. C. Mitani, and L. Vigilant, "The Limited Impact of Kinship on Cooperation in Wild Chimpanzees," *Proceedings of the National Academy of Sciences of the United States of America* 104, no. 19 (2007): 7786–7790.

51. Goodall, *Chimpanzees of Gombe*, 529.

52. Ibid., 532.

53. Richard W. Wrangham and Dale Peterson, *Demonic Males: Apes and the Origins of Human Violence* (Boston: Houghton Mifflin, 1996).

54. G. G. Gallup et al., "Further Reflections on Self-Recognition in Primates," *Animal Behaviour* 50 (1995): 1525–1532.

55. Tomasello and Call, *Primate Cognition*, 331.

56. C. M. Heyes, "Self-Recognition in Primates: Further Reflections Create a Hall of Mirrors," *Animal Behaviour* 50 (1995): 1533–1542.

57. S. M. Reader and K. N. Laland, "Social Intelligence, Innovation, and Enhanced Brain Size in Primates," *Proceedings of the National Academy of Sciences of the United States of America* 99, no. 7 (2002): 4436–4441.

58. Ronald George Pearson, *The Avian Brain* (London: Academic, 1972).

59. N. J. Emery, "Cognitive Ornithology: The Evolution of Avian Intelligence," *Philosophical Transactions of the Royal Society B—Biological Sciences* 361, no. 1465 (2006): 23–43.

60. N. J. Emery and N. S. Clayton, "Comparing the Complex Cognition of Birds and Primates," in *Comparative Vertebrate Cognition: Are Primates Superior to Non-Primates?* ed. Lesley J. Rogers and Gisela T. Kaplan (New York: Kluwer Academic/Plenum, 2004), 3–55.

61. Ibid., 6–7.

62. Ibid., 8–9; A. Reiner, K. Yamamoto, and H. J. Karten, "Organization and Evolution of the Avian Forebrain," *Anatomical Record Part A—Discoveries in Molecular Cellular and Evolutionary Biology* 287A, no. 1 (2005): 1080–1102.

63. E. D. Jarvis et al., "Avian Brains and a New Understanding of Vertebrate Brain Evolution," *Nature Reviews Neuroscience* 6, no. 2 (2005): 151–159.

64. M. J. Burish, H. Y. Kueh, and S. S. H. Wang, "Brain Architecture and Social Complexity in Modern and Ancient Birds," *Brain Behavior and Evolution* 63, no. 2 (2004): 107–124; A. N. Iwaniuk and J. E. Nelson, "Developmental Differences Are Correlated with Relative Brain Size in Birds: A Comparative Analysis," *Canadian Journal of Zoology—Revue Canadienne De Zoologie* 81, no. 12 (2003): 1913–1928.

65. Iwaniuk and Nelson, "Developmental Differences Are Correlated with Relative Brain Size in Birds," 1925.

66. T. Bugnyar and K. Kotrschal, "Observational Learning and the Raiding of Food Caches in Ravens, *Corvus Corax*: Is It 'Tactical' Deception?" *Animal Behaviour* 64 (2002): 185–195.

67. Bernd Heinrich, *Mind of the Raven: Investigations and Adventures with Wolf-Birds* (New York: Cliff Street, 1999).

68. Ibid., 139.

69. Ibid., 174–176.

70. Ibid., 273.

71. Ibid., iii.

72. Ibid., 234.

73. Bernd Heinrich, *Ravens in Winter* (New York: Summit, 1989).

74. Heinrich, *Mind of the Raven*, 244.

75. A. Seed, N. Emery, and N. Clayton. "Intelligence in Corvids and Apes: A Case of Convergent Evolution?" *Ethology* 115, no. 5 (2009): 401–420.

76. A. M. Seed, N. S. Clayton, and N. J. Emery. "Postconflict Third-Party Affiliation in Rooks, *Corvus Frugilegus*," *Current Biology* 17, no. 2 (2007): 152–158.

77. Heinrich, *Mind of the Raven*, 261.

78. Bugnyar and Kotrschal, "Observational Learning," 74.

79. T. Bugnyar and B. Heinrich, "Ravens, *Corvus corax*, Differentiate Between Knowledgeable and Ignorant Competitors," *Proceedings of the Royal Society B—Biological Sciences* 272, no. 1573 (2005): 1641–1646.

80. Hare, Call, and Tomasello, "Chimpanzees Deceive a Human Competitor by Hiding," 495.

81. N. J. Emery and N. S. Clayton, "Effects of Experience and Social Context on Prospective Caching Strategies by Scrub Jays," *Nature* 414, no. 6862 (2001): 443–446.

82. Ibid., 445.

83. Heinrich, *Mind of the Raven*, 196; D. A. Ratcliffe, *The Raven: A Natural History in Britain and Ireland* (London: Poyser, 1997), 266.

84. R. N. Conner, "Vocalizations of Common Ravens in Virginia," *Condor* 87, no. 3 (1985): 379–388.

85. Heinrich, *Mind of the Raven*, 196.

86. Heinrich, *Ravens in Winter*, 252.

87. Heinrich, *Mind of the Raven*, 191.

88. Heinrich, *Ravens in Winter*, 249.

89. Heinrich, *Mind of the Raven*, 198.

90. B. Heinrich, "Testing Insight in Ravens," in *The Evolution of Cognition*, ed. Cecilia M. Heyes and Ludwig Huber (Cambridge, Mass.: MIT Press, 2000), 289–305.

91. Heinrich, *Mind of the Raven*, 137–139.

92. Ibid., 71.

93. N. J. Emery and N. S. Clayton, "The Mentality of Crows: Convergent Evolution of Intelligence in Corvids and Apes," *Science* 306, no. 5703 (2004): 1903–1907; J. A. Basil et al., "Differences in Hippocampal Volume Among Food Storing Corvids," *Brain Behavior and Evolution* 47, no. 3 (1996): 156–164.

94. N. S. Clayton and A. Dickinson, "Episodic-Like Memory During Cache Recovery by Scrub Jays," *Nature* 395, no. 6699 (1998): 272–274.

95. L. Lefebvre, N. Nicolakakis, and D. Boire, "Tools and Brains in Birds," *Behaviour* 139 (2002): 939–973.

96. G. R. Hunt and R. D. Gray, "The Crafting of Hook Tools by Wild New Caledonian Crows," *Proceedings of the Royal Society of London Series B—Biological Sciences* 271 (2004): S88–S90.

97. G. R. Hunt, "Human-Like, Population-Level Specialization in the Manufacture of Pandanus Tools by New Caledonian Crows *Corvus moneduloides*," *Proceedings of the Royal Society of London Series B—Biological Sciences* 267, no. 1441 (2000): 403–413.

98. Ibid., 407.

99. A. A. S. Weir, J. Chappell, and A. Kacelnik, "Shaping of Hooks in New Caledonian Crows," *Science* 297, no. 5583 (2002): 981.

100. C. D. Bird and N. J. Emery, "Insightful Problem Solving and Creative Tool Modification by Captive Nontool-Using Rooks," *Proceedings of the National Academy of Sciences of the United States of America* (2009): 10370–10375.

101. Rumbaugh and Washburn, *Intelligence of Apes*, 133.

102. A. H. Taylor et al., "Spontaneous Metatool Use by New Caledonian Crows," *Current Biology* 17, no. 17 (2007): 1504–1507.

103. J. Cnotka et al., "Extraordinary Large Brains in Tool-Using New Caledonian Crows (*Corvus moneduloides*)," *Neuroscience Letters* 433, no. 3 (2008): 241–245.

104. Heinrich, "Testing Insight in Ravens," 298.

105. Ratcliffe, *Raven*, 253.

106. Heinrich, *Mind of the Raven*, 306–307.

107. Ibid., 284–291; Ratcliffe, *Raven*, 116.

108. Heinrich, *Mind of the Raven*, 137.

109. Ibid., 172.

110. H. Prior, A. Schwarz, and O. Gunturkun, "Mirror-Induced Behavior in the Magpie (*Pica pica*): Evidence of Self-Recognition," *PLoS Biology* 6, no. 8 (2008): 1642–1650.

111. Georg F. Striedter, *Principles of Brain Evolution* (Sunderland, Mass.: Sinauer, 2005).

112. Emery and Clayton, "Mentality of Crows," 1903–1907; L. Marino, "Big Brains Do Matter in New Environments," *Proceedings of the National Academy of Sciences of the United States of America* 102, no. 15 (2005): 5306–5307.

6. The Evolution of Personality

1. Joyce H. Poole and Cynthia J. Moss, "Elephant Sociality and Complexity: The Scientific Evidence," in *Elephants and Ethics: Toward a Morality of Coexistence*, ed. Christen M. Wemmer and Catherine A. Christen (Baltimore: Johns Hopkins University Press, 2008), 69–98.

2. S. Herculano-Houzel, "Brains Matter, Bodies Maybe Not: The Case for Examining Neuron Numbers Irrespective of Body Size," *Annals of the New York Academy of Sciences* 1225, no. 1 (2011): 191–199.

3. A. Y. Hakeem et al., "Von Economo Neurons in the Elephant Brain," *Anatomical Record—Advances in Integrative Anatomy and Evolutionary Biology* 292, no. 2 (2009): 242–248.

4. A. Y. Hakeem et al., "Brain of the African Elephant (*Loxodonta Africana*): Neuroanatomy from Magnetic Resonance Images," *Anatomical Record Part A—Discoveries in Molecular, Cellular, and Evolutionary Biology* 287, no. 1 (2005): 1117–1127.

5. P. C. Lee and C. J. Moss, "The Social Context of Learning and Behavioural Development Among Wild African Elephants," in *Mammalian Social Learning: Comparative and Ecological Perspectives*, ed. Hilary O. Box and Kathleen Rita Gibson (Cambridge: Cambridge University Press, 1999), 102–125; K. Payne, "Sources of Social Complexity in the Three Elephant Species," in *Animal Social Complexity: Intel-*

ligence, Culture, and Individualized Societies, ed. F. B. M. de Waal and Peter L. Tyack (Cambridge, Mass.: Harvard University Press, 2003), 57–85.

6. L. S. Eggert, C. A. Rasner, and D. S. Woodruff, "The Evolution and Phylogeography of the African Elephant Inferred from Mitochondrial DNA Sequence and Nuclear Microsatellite Markers," *Proceedings of the Royal Society of London Series B—Biological Sciences* 269, no. 1504 (2002): 1993–2006.

7. R. Sukumar, *The Living Elephants: Evolutionary Ecology, Behavior, and Conservation* (New York: Oxford University Press, 2003).

8. Cynthia Moss, *Elephant Memories: Thirteen Years in the Life of an Elephant Family* (New York: Morrow, 1988).

9. Poole and Moss, "Elephant Sociality and Complexity," 75.

10. G. Wittemyer, I. Douglas-Hamilton, and W. M. Getz, "The Socioecology of Elephants: Analysis of the Processes Creating Multitiered Social Structures," *Animal Behaviour* 69 (2005): 1357–1371.

11. E. A. Archie, C. J. Moss, and S. C. Alberts, "The Ties That Bind: Genetic Relatedness Predicts the Fission and Fusion of Social Groups in Wild African Elephants," *Proceedings of the Royal Society B—Biological Sciences* 273, no. 1586 (2006): 513–522.

12. Ibid., 517–520.

13. Moss, *Elephant Memories*, 219.

14. K. Mccomb et al., "Unusually Extensive Networks of Vocal Recognition in African Elephants," *Animal Behaviour* 59 (2000): 1103–1109.

15. Wittemyer, Douglas-Hamilton, and Getz, "Socioecology of Elephants," 1367.

16. K. Mccomb et al., "Matriarchs as Repositories of Social Knowledge in African Elephants," *Science* 292, no. 5516 (2001): 491–494.

17. I. Douglas-Hamilton et al., "Behavioural Reactions of Elephants Towards a Dying and Deceased Matriarch," *Applied Animal Behaviour Science* 100, no. 1-2 (2006): 87–102.

18. Sukumar, *Living Elephants*, 141–142.

19. L. A. Bates, J. H. Poole, and R. W. Byrne, "Elephant Cognition," *Current Biology* 18, no. 13 (2008): R544–R546.

20. J. H. Poole et al., "Elephants Are Capable of Vocal Learning," *Nature* 434, no. 7032 (2005): 455–456.

21. Ibid., 455.

22. J. H. Shoshani, "It's a Nose! It's a Hand! It's an Elephant Trunk!" *Natural History* 106, no. 10 (1997): 36–45.

23. Payne, "Sources of Social Complexity," 76; J. H. Poole, "Signals and Assessment in African Elephants: Evidence from Playback Experiments," *Animal Behaviour* 58 (1999): 185–193.

24. Mccomb et al., "Unusually Extensive Networks of Vocal Recognition," 1107; Sukumar, *Living Elephants*, 147.

25. Sukumar, *Living Elephants*, 149.

26. Ibid., 152–155.

27. Bates, Poole, and Byrne, "Elephant Cognition," R544–R546.

28. Shoshani, "It's a Nose!" 36–45.

29. Ibid., 40.

30. Charles Darwin, *The Descent of Man, and Selection in Relation to Sex* (London: Murray, 1871); B. L. Hart et al., "Cognitive Behaviour in Asian Elephants: Use and Modification of Branches for Fly Switching," *Animal Behaviour* 62 (2001): 839–847.

31. Hart et al., "Cognitive Behaviour in Asian Elephants," 845.

32. Poole and Moss, "Elephant Sociality and Complexity," 88–89.

33. Sukumar, *Living Elephants*, 27.

34. Shoshani, "It's a Nose!" 36–45.

35. Sukumar, *Living Elephants*, 177.

36. Shoshani, "It's a Nose!" 43.

37. Moss, *Elephant Memories*, 128.

38. Ibid., 73–74.

39. Douglas-Hamilton et al., "Behavioural Reactions of Elephants," 88.

40. Payne, "Sources of Social Complexity," 81–82.

41. Moss, *Elephant Memories*, 270.

42. K. Mccomb, L. Baker, and C. Moss, "African Elephants Show High Levels of Interest in the Skulls and Ivory of Their Own Species," *Biology Letters* 2, no. 1 (2006): 26–28.

43. Mccomb et al., "Unusually Extensive Networks of Vocal Recognition," 1103–1109.

44. Payne, "Sources of Social Complexity," 79–80.

45. Lee and Moss, "Social Context of Learning and Behavioural Development," 102–125.

46. Moss, *Elephant Memories*, 265.

47. Lee and Moss, "Social Context of Learning and Behavioural Development," 110–111.

48. G. A. Bradshaw et al., "Elephant Breakdown," *Nature* 433, no. 7028 (2005): 807–807.

49. J. M. Plotnik, F. B. M. de Waal, and D. Reiss, "Self-Recognition in an Asian Elephant," *Proceedings of the National Academy of Sciences of the United States of America* 103, no. 45 (2006): 17053–17057.

50. B. L. Hart, L. A. Hart, and N. Pinter-Wollman, "Large Brains and Cognition: Where Do Elephants Fit In?" *Neuroscience and Biobehavioral Reviews* 32, no. 1 (2008): 86–98.

51. L. Marino, "A Comparison of Encephalization Between Odontocete Cetaceans and Anthropoid Primates," *Brain Behavior and Evolution* 51, no. 4 (1998): 230–238.

52. S. Herculano-Houzel, "The Human Brain in Numbers: A Linearly Scaled-up Primate Brain," *Frontiers in Human Neuroscience* 3 (2009): article 31.

53. A. C. Tshudin, "Belief Attribution Tasks with Dolphins: What Social Minds Can Reveal About Animal Rationality," in *Rational Animals?* ed. S. L. Hurley and Matthew Nudds (Oxford: Oxford University Press, 2006), 413–436.

54. L. Marino, "Convergence of Complex Cognitive Abilities in Cetaceans and Primates," *Brain Behavior and Evolution* 59, no. 1-2 (2002): 21–32.

55. P. R. Hof, R. Chanis, and L. Marino, "Cortical Complexity in Cetacean Brains," *Anatomical Record Part A—Discoveries in Molecular Cellular and Evolutionary Biology* 287A, no. 1 (2005): 1142–1152.

56. C. Butti et al., "Total Number and Volume of Von Economo Neurons in the Cerebral Cortex of Cetaceans," *Journal of Comparative Neurology* 515, no. 2 (2009): 243–259.

57. L. M. Herman, "Intelligence and Rational Behaviour in the Bottlenosed Dolphin," in *Rational Animals?* ed. Hurley and Nudds, 239–267.

58. L. Marino, D. W. McShea, and M. D. Uhen, "Origin and Evolution of Large Brains in Toothed Whales," *Anatomical Record Part A—Discoveries in Molecular Cellular and Evolutionary Biology* 281A, no. 2 (2004): 1247–1255.

59. R. C. Connor, M. R. Heithaus, and L. M. Barre, "Complex Social Structure, Alliance Stability and Mating Access in a Bottlenose Dolphin 'Super-Alliance,'" *Proceedings of the Royal Society of London Series B—Biological Sciences* 268, no. 1464 (2001): 263–267; R. C. Connor, "Dolphin Social Intelligence: Complex Alliance Relationships in Bottlenose Dolphins and a Consideration of Selective Environments for Extreme Brain Size Evolution in Mammals," *Philosophical Transactions of the Royal Society B—Biological Sciences* 362, no. 1480 (2007): 587–602.

60. Connor, Heithaus, and Barre, "Complex Social Structure," 263.

61. C. M. Johnson and K. S. Norris, "Delphinid Social Organization and Social Behavior," in *Dolphin Cognition and Behavior: A Comparative Approach*, ed. Ronald J. Schusterman, Jeanette A. Thomas, and Forrest G. Wood (Hillsdale, N.J.: Erlbaum, 1986), 335–346.

62. R. Smolker, "Keeping in Touch at Sea: Group Movement in Dolphins and Whales," in *On the Move: How and Why Animals Travel in Groups*, ed. Sue Boinski and Paul Alan Garber (Chicago: University of Chicago Press, 2000), 559–586.

63. V. M. Bel'kovich, "Herd Structure, Hunting, and Play: Bottlenose Dolphins in the Black Sea," in *Dolphin Societies: Discoveries and Puzzles*, ed. Karen Pryor and Kenneth S. Norris (Berkeley: University of California Press, 1991), 17–78.

64. Smolker, "Keeping in Touch at Sea," 574.

65. B. L. Sargeant et al., "Specialization and Development of Beach Hunting, a Rare Foraging Behavior, by Wild Bottlenose Dolphins (*Tursiops sp.*)," *Canadian Journal of Zoology—Revue Canadienne De Zoologie* 83, no. 11 (2005): 1400–1410.

66. R. S. Wells, D. J. Boness, and G. B. Rathbun, "Behavior," in *Biology of Marine Mammals*, ed. John Elliott Reynolds and Sentiel A. Rommel (Washington D.C.: Smithsonian Institution, 1999), 324–422.

67. J. Mann et al., "Why Do Dolphins Carry Sponges?" *PLoS ONE* 3, no. 12 (2008): 1–7.

68. B. L. Sargeant and J. Mann, "Developmental Evidence for Foraging Traditions in Wild Bottlenose Dolphins," *Animal Behaviour* 78, no. 3 (2009): 715–721.

69. C. E. Bender, D. L. Herzing, and D. F. Bjorklund, "Evidence of Teaching in Atlantic Spotted Dolphins (*Stenella frontalis*) by Mother Dolphins Foraging in the Presence of Their Calves," *Animal Cognition* 12, no. 1 (2009): 43–53; T. M. Caro and M. D. Hauser, "Is There Teaching in Nonhuman Animals?" *Quarterly Review of Biology* 67, no. 2 (1992): 151–174.

70. Mann, "Why Do Dolphins Carry Sponges?" 1.

71. J. Finn, T. Tregenza, and M. Norman, "Preparing the Perfect Cuttlefish Meal: Complex Prey Handling by Dolphins," *PLoS One* 4, no. 1 (2009): e4217.

72. Sargeant and Mann, "Developmental Evidence for Foraging Traditions," 715–716; Wells, Boness, and Rathbun, "Behavior," 324–422; B. Wursig, "Delphinid Foraging Strategies," in *Dolphin Cognition and Behavior*, ed. Schusterman, Thomas, and Wood, 347–359.

73. P. L. Tyack, "Communication and Cognition," in *Biology of Marine Mammals*, ed. Reynolds and Rommel, 287–323.

74. Herman, "Intelligence and Rational Behaviour," 454–455.

75. Donald R. Griffin, *Animal Minds: Beyond Cognition to Consciousness* (Chicago: University of Chicago Press, 2001).

76. Tyack, "Communication and Cognition," 287–323.

77. Herman, "Intelligence and Rational Behaviour," 454.

78. Smolker, "Keeping in Touch at Sea," 580.

79. V. M. Janik, "Whistle Matching in Wild Bottlenose Dolphins (*Tursiops truncatus*)," *Science* 289, no. 5483 (2000): 1355–1357.

80. Ibid., 1355.

81. P. Tyack, "If You Need Me, Whistle," *Natural History* 100 (1991): 60–61.

82. V. M. Janik, L. S. Sayigh, and R. S. Wells, "Signature Whistle Shape Conveys Identity Information to Bottlenose Dolphins," *Proceedings of the National Academy of Sciences of the United States of America* 103, no. 21 (2006): 8293–8297.

83. R. S. Wells, "Bringing Up Baby," *Natural History* 100, no. 8 (1991): 56–62.

84. D. G. Richards, "Dolphin Vocal Mimicry and Vocal Object Labeling," in *Dolphin Cognition and Behavior*, ed. Schusterman, Thomas, and Wood, 273–288.

85. Smolker, "Keeping in Touch at Sea," 578.

86. L. M. Herman, "Cognition and Language Competencies in Bottlenosed Dolphins," in *Dolphin Cognition and Behavior*, ed. Schusterman, Thomas, and Wood, 221–252.

87. Herman, "Intelligence and Rational Behaviour," 458.

88. L. M. Herman, "In Which Procrustean Bed Does the Sea Lion Sleep Tonight?" *Psychological Record* 39, no. 1 (1989): 19–50.

89. Herman, "Intelligence and Rational Behaviour," 447.

90. Herman, "Cognition and Language Competencies," 238–239.

91. L. Marino et al., "Cetaceans Have Complex Brains for Complex Cognition," *PLoS Biology* 5, no. 5 (2007): 966–972.

92. R. J. Schusterman and R. C. Gisiner, "Pinnipeds, Porpoises, and Parsimony: Animal Language Research Viewed from a Bottom-up Perspective," in *Anthropomorphism, Anecdotes, and Animals*, ed. Robert W. Mitchell, Nicholas S. Thompson, and H. Lyn Miles (Albany: State University of New York Press, 1997), 370–382.

93. G. A. J. Worthy and J. P. Hickie, "Relative Brain Size in Marine Mammals," *American Naturalist* 128, no. 4 (1986): 445–459.

94. R. J. Schusterman and R. Gisiner, "Artificial Language Comprehension in Dolphins and Sea Lions—The Essential Cognitive Skills," *Psychological Record* 38, no. 3 (1988): 311–348.

95. R. C. Connor and K. S. Norris, "Are Dolphins Reciprocal Altruists?" *American Naturalist* 119, no. 3 (1982): 358–374.

96. Ibid., 367–369.

97. John Elliott Reynolds, Randall S. Wells, and Samantha D. Eide, *The Bottlenose Dolphin: Biology and Conservation* (Gainesville: University Press of Florida, 2000).

98. Wells, "Bringing Up Baby," 56–62.

99. Connor and Norris, "Are Dolphins Reciprocal Altruists?" 360–372.

100. Wells, "Bringing Up Baby," 56–62.

101. Connor and Norris, "Are Dolphins Reciprocal Altruists?" 371.

102. M. C. Caldwell and D. K. Caldwell, "Epimeletic (Caregiving) Behavior in Cetacea," in *Whales, Dolphins, and Porpoises: (Proceedings)*, ed. Kenneth S. Norris (Berkeley: University of California Press, 1966), 755–789.

103. Ibid., 767.

104. Connor and Norris, "Are Dolphins Reciprocal Altruists?" 372.

105. D. Reiss and L. Marino, "Mirror Self-Recognition in the Bottlenose Dolphin: A Case of Cognitive Convergence," *Proceedings of the National Academy of Sciences of the United States of America* 98, no. 10 (2001): 5937–5942.

106. Butti et al., "Total Number and Volume of Von Economo Neurons," 255–257.

107. Marino, "Convergence of Complex Cognitive Abilities," 29.

108. Ibid., 29.

7. Concepts as Feature Extraction

1. S. Herculano-Houzel, "Brains Matter, Bodies Maybe Not: The Case for Examining Neuron Numbers Irrespective of Body Size," *Annals of the New York Academy of Sciences* 1225, no. 1 (2011): 191–199; Harry J. Jerison, *Evolution of the Brain and Intelligence* (New York: Academic, 1973).

2. Dee Unglaub Silverthorn, *Human Physiology: An Integrated Approach*, 3rd ed. (San Francisco: Pearson Benjamin Cummings, 2004).

3. E. A. Rossi and A. Roorda, "The Relationship Between Visual Resolution and Cone Spacing in the Human Fovea," *Nature Neuroscience* 13, no. 2 (2010): 156–157; N. Drasdo, C. L. Millican, C. R. Katholi, and C. A. Curcio, "The Length of Henle Fibers in the Human Retina and a Model of Ganglion Receptive Field Density in the Visual Field," *Vision Research* 47, no. 22 (2007): 2901–2911.

4. B. Volgyi et al., "Convergence and Segregation of the Multiple Rod Pathways in Mammalian Retina," *Journal of Neuroscience* 24, no. 49 (2004): 11182–11192.

5. J. Elstrott and M. B. Feller, "Vision and the Establishment of Direction-Selectivity: A Tale of Two Circuits," *Current Opinion in Neurobiology* 19 (2009): 293–297; Z. J. Zhou and S. Lee, "Synaptic Physiology of Direction Selectivity in the Retina," *Journal of Physiology* 586, part 18 (2008): 4371–4376.

6. Dale Purves, *Neuroscience*, 4th ed. (Sunderland, Mass.: Sinauer, 2008).

7. T. Serre, A. Oliva, and T. Poggio, "A Feedforward Architecture Accounts for Rapid Categorization," *Proceedings of the National Academy of Sciences of the United States of America* 104, no. 15 (2007): 6424–6429.

8. A. Pasupathy and C. E. Connor, "Population Coding of Shape in Area V4," *Nature Neuroscience* 5, no. 12 (2002): 1332–1338.

9. Christof Koch, *The Quest for Consciousness: A Neurobiological Approach* (Denver: Roberts, 2004).

10. Semir Zeki, *Splendors and Miseries of the Brain: Love, Creativity, and the Quest for Human Happiness* (Chichester, U.K.: Wiley-Blackwell, 2009).

11. Ibid., 30–31.

12. J. M. Fuster, "Frontal Lobe and Cognitive Development," *Journal of Neurocytology* 31, no. 3-5 (2002): 373–385.

13. Michael S. Gazzaniga, *Human: The Science Behind What Makes Us Unique* (New York: Ecco, 2008).

14. E. K. Miller and T. J. Buschman, "Rules Through Recursion: How Interactions Between the Frontal Cortex and Basal Ganglia May Build Abstract, Complex Rules from Concrete, Simple Ones," in *Neuroscience of Rule-Guided Behavior*, ed. Silvia A. Bunge and Jonathan D. Wallis (Oxford: Oxford University Press, 2008), 419–440; D. J. Freedman, "Exploring the Roles of the Frontal, Temporal, and Parietal Lobes in Visual Categorization," in *Neuroscience of Rule-Guided Behavior*, ed. Bunge and Wallis, 391–418.

15. N. Gogtay et al., "Dynamic Mapping of Human Cortical Development During Childhood Through Early Adulthood," *Proceedings of the National Academy of Sciences of the United States of America* 101, no. 21 (2004): 8174–8179.

16. T. M. Preuss, "Evolutionary Specializations of Primate Brain Systems," in *Primate Origins: Adaptations and Evolution*, ed. Matthew J. Ravosa and Marian Dagosto (New York: Springer, 2007), 625–675.

17. Fuster, "Frontal Lobe and Cognitive Development," 376.

18. P. T. Schoenemann, M. J. Sheehan, and L. D. Glotzer, "Prefrontal White Matter Volume Is Disproportionately Larger in Humans than in Other Primates," *Nature Neuroscience* 8, no. 2 (2005): 242–252.

19. K. Semendeferi et al., "Prefrontal Cortex in Humans and Apes: A Comparative Study of Area 10," *American Journal of Physical Anthropology* 114, no. 3 (2001): 224–241; R. L. Holloway, "Brief Communication: How Much Larger Is the Relative Volume of Area 10 of the Prefrontal Cortex in Humans?" *American Journal of Physical Anthropology* 118, no. 4 (2002): 399–401.

20. N. Ramnani and A. M. Owen, "Anterior Prefrontal Cortex: Insights into Function from Anatomy and Neuroimaging," *Nature Reviews Neuroscience* 5, no. 3 (2004): 184–194.

21. K. Christoff and K. Keramatian, "Abstraction of Mental Representations: Theoretical Considerations and Neuroscientific Evidence," in *Neuroscience of Rule-Guided Behavior*, ed. Bunge and Wallis, 107–126.

22. A. E. Green et al., "Frontopolar Cortex Mediates Abstract Integration in Analogy," *Brain Research* 1096 (2006): 125–137.

23. A. E. Green et al., "Connecting Long Distance: Semantic Distance in Analogical Reasoning Modulates Frontopolar Cortex Activity," *Cerebral Cortex* 20, no. 1 (2009): 70–76.

24. Keith James Holyoak and Paul Thagard, *Mental Leaps: Analogy in Creative Thought* (Cambridge, Mass.: MIT Press, 1995).

25. D. Badre et al., "Hierarchical Cognitive Control Deficits Following Damage to the Human Frontal Lobe," *Nature Neuroscience* 12, no. 4 (2009): 515–522; K. Christoff et al., "Prefrontal Organization of Cognitive Control According to Levels of Abstraction," *Brain Research* 1286 (2009): 94–105.

26. W. W. Seeley et al., "Early Frontotemporal Dementia Targets Neurons Unique to Apes and Humans," *Annals of Neurology* 60, no. 6 (2006): 660–667.

27. A. D. Craig, "How Do You Feel—Now? The Anterior Insula and Human Awareness," *Nature Reviews Neuroscience* 10, no. 1 (2009): 59–70.

28. J. M. Allman et al., "Intuition and Autism: A Possible Role for Von Economo Neurons," *Trends in Cognitive Sciences* 9, no. 8 (2005): 367–373.

29. C. Fajardo et al., "Von Economo Neurons Are Present in the Dorsolateral (Dysgranular) Prefrontal Cortex of Humans," *Neuroscience Letters* 435, no. 3 (2008): 215–218.

30. Seeley et al., "Early Frontotemporal Dementia," 663.

31. J. Allman and A. Hasenstaub, "Brains, Maturation Times, and Parenting," *Neurobiology of Aging* 20, no. 4 (1999): 447–454; Allman et al., "Intuition and Autism," 453.

32. Georg F. Striedter, *Principles of Brain Evolution* (Sunderland, Mass.: Sinauer, 2005).

33. Ibid., 205.

34. Ibid., 143.

35. D. C. VanEssen, "A Tension-Based Theory of Morphogenesis and Compact Wiring in the Central Nervous System," *Nature* 385, no. 6614 (1997): 313–318.

36. S. Herculano-Houzel et al., "Connectivity-Driven White Matter Scaling and Folding in Primate Cerebral Cortex," *Proceedings of the National Academy of Sciences of the United States of America* 107, no. 44 (2010): 19008–19013; Striedter, *Principles of Brain Evolution*, 128–134.

37. VanEssen, "Tension-Based Theory of Morphogenesis and Compact Wiring," 313–315.

38. A. S. Chiang et al., "Three-Dimensional Reconstruction of Brain-Wide Wiring Networks in Drosophila at Single-Cell Resolution," *Current Biology* 21 (2011): 1–11.

39. O. Sporns, G. Tononi, and R. Kotter, "The Human Connectome: A Structural Description of the Human Brain," *PLoS Computational Biology* 1, no. 4 (2005): e42; V. B. Mountcastle, "The Columnar Organization of the Neocortex," *Brain* 120 (part 4) (1997): 701–722; D. P. Buxhoeveden and M. F. Casanova, "The Minicolumn and Evolution of the Brain," *Brain, Behavior and Evolution* 60, no. 3 (2002): 125–151.

40. Striedter, *Principles of Brain Evolution*, 286, 334.

41. F. A. Azevedo et al., "Equal Numbers of Neuronal and Nonneuronal Cells Make the Human Brain an Isometrically Scaled-up Primate Brain," *Journal of Comparative Neurology* 513, no. 5 (2009): 532–541.

42. S. Herculano-Houzel, "The Human Brain in Numbers: A Linearly Scaled-up Primate Brain," *Frontiers in Human Neuroscience* 3 (2009): 31.

43. S. Herculano-Houzel, "Coordinated Scaling of Cortical and Cerebellar Numbers of Neurons," *Frontiers in Neuroanatomy* 4 (2010): 1–8.

44. Herculano-Houzel, "Brains Matter, Bodies Maybe Not," 195.

45. Herculano-Houzel, "Human Brain in Numbers."

46. J. H. Balsters et al., "Evolution of the Cerebellar Cortex: The Selective Expansion of Prefrontal-Projecting Cerebellar Lobules," *Neuroimage* 49, no. 3 (2010): 2045–2052.

47. C. E. MacLeod et al., "Expansion of the Neocerebellum in Hominoidea," *Journal of Human Evolution* 44, no. 4 (2003): 401–429; J. B. Smaers, J. Steele, and K. Zilles, "Modeling the Evolution of Cortico-Cerebellar Systems in Primates," *Annals of the New York Academy of Sciences* 1225, no. 1 (2011): 176–190.

48. Herculano-Houzel, "Human Brain in Numbers," 10.

49. N. J. Emery and N. S. Clayton, "The Mentality of Crows: Convergent Evolution of Intelligence in Corvids and Apes," *Science* 306, no. 5703 (2004): 1903–1907.

50. J. Cnotka et al., "Extraordinary Large Brains in Tool-Using New Caledonian Crows (*Corvus moneduloides*)," *Neuroscience Letters* 433, no. 3 (2008): 241–245.

51. L. Chittka and J. Niven, "Are Bigger Brains Better?" *Current Biology* 19, no. 21 (2009): R995–R1008; S. M. Farris, "Structural, Functional and Developmental Convergence of the Insect Mushroom Bodies with Higher Brain Centers of Vertebrates," *Brain, Behavior and Evolution* 72, no. 1 (2008): 1–15.

52. Farris, "Structural, Functional and Developmental Convergence," 12.

53. C. E. Collins et al., "Neuron Densities Vary Across and Within Cortical Areas in Primates," *Proceedings of the National Academy of Sciences of the United States of America* 107, no. 36 (2010): 15927–15932.

54. S. Herculano-Houzel, "Scaling of Brain Metabolism with a Fixed Energy Budget per Neuron: Implications for Neuronal Activity, Plasticity and Evolution," *PLoS One* 6, no. 3 (2011): e17514.

55. J. Karbowski, "Global and Regional Brain Metabolic Scaling and Its Functional Consequences," *BMC Biology* 5, no. 18 (2007).

56. S. M. Fleming et al., "Relating Introspective Accuracy to Individual Differences in Brain Structure," *Science* 329, no. 5998 (2010): 1541–1543.

57. Striedter, *Principles of Brain Evolution*, 356–359.

58. Gazzaniga, *Human*, 14–15; Katherine S. Pollard, "What Makes Us Human?" *Scientific American* 300, no. 5 (2009): 44–49.

59. D. Bishop, "Genes, Cognition, and Communication Insights from Neurodevelopmental Disorders," *Year in Cognitive Neuroscience 2009* 1156 (2009): 1–18.

60. M. Caceres et al., "Elevated Gene Expression Levels Distinguish Human from Non-Human Primate Brains," *Proceedings of the National Academy of Sciences of the United States of America* 100, no. 22 (2003): 13030–13035.

8. The Brain and Belief

1. O. Bar-Yosef, "The Upper Paleolithic Revolution," *Annual Review of Anthropology* 31 (2002): 363–393; S. Mithen, "Ethnobiology and the Evolution of the Human Mind," *Journal of the Royal Anthropological Institute* 12 (2006): S45–S61; Ian Tattersall, *The Monkey in the Mirror: Essays on Science and What Makes Us Human* (New York: Harcourt, 2002); J. F. Hoffecker, "Innovation and Technological Knowledge in the Upper Paleolithic of Northern Eurasia," *Evolutionary Anthropology* 14, no. 5 (2005): 186–198.

2. I. Tattersall, "The Dual Origin of Modern Humanity," *Collegium Antropologicum* 28 (2004): 77–85; Brian Hayden, *Shamans, Sorcerers, and Saints: A Prehistory of Religion* (Washington, D.C.: Smithsonian Books, 2003).

3. A. Nowell, "From a Paleolithic Art to Pleistocene Visual Cultures (Introduction to Two Special Issues on 'Advances in the Study of Pleistocene Imagery and Symbol Use')," *Journal of Archaeological Method and Theory* 13, no. 4 (2006): 239–249.

4. Steven J. Mithen, *The Prehistory of the Mind: A Search for the Origins of Art, Religion, and Science* (London: Thames and Hudson, 1996).

5. Ibid., 195; Mithen, "Ethnobiology and the Evolution of the Human Mind," S51.

6. J. David Lewis-Williams, *The Mind in the Cave: Consciousness and the Origins of Art* (London: Thames and Hudson, 2002).

7. Ibid., 80.

8. F. d'Errico et al., "Archaeological Evidence for the Emergence of Language, Symbolism, and Music—An Alternative Multidisciplinary Perspective," *Journal of World Prehistory* 17, no. 1 (2003): 1–70.

9. Bar-Yosef, "Upper Paleolithic Revolution," 378.

10. Jean Clottes and J. David Lewis-Williams, *The Shamans of Prehistory: Trance and Magic in the Painted Caves* (New York: Abrams, 1998).

11. Hayden, *Shamans, Sorcerers, and Saints*, 136.

12. Clottes and Lewis-Williams, *Shamans of Prehistory*, 110.

13. Ibid., 86–91.

14. Lewis-Williams, *Mind in the Cave*, 139.

15. I. M. Lewis, *Ecstatic Religion: A Study of Shamanism and Spirit Possession*, 2nd ed. (London: Routledge, 1989).

16. Lewis-Williams, *Mind in the Cave*, 167–170.

17. Scott Atran, *In Gods We Trust: The Evolutionary Landscape of Religion, Evolution and Cognition* (Oxford: Oxford University Press, 2002).

18. Mithen, "Ethnobiology and the Evolution of the Human Mind," S47.

19. Hoffecker, "Innovation and Technological Knowledge in the Upper Paleolithic, 188; Mithen, "Ethnobiology and the Evolution of the Human Mind," S47.

20. Tattersall, "Dual Origin of Modern Humanity," 82.

21. Lewis-Williams, *Mind in the Cave*, 74.

22. Ibid., 170–171, 215.

23. Michael Winkelman, *Shamanism: The Neural Ecology of Consciousness and Healing* (Westport, Conn.: Bergin and Garvey, 2000).

24. Quoted in *"Up* Series," *Wikipedia*, http://en.wikipedia.org/wiki/Up_Series (accessed June 5, 2011).

25. C. S. Alcorta and R. Sosis, "Ritual, Emotion, and Sacred Symbols—The Evolution of Religion as an Adaptive Complex," *Human Nature—An Interdisciplinary Biosocial Perspective* 16, no. 4 (2005): 323–359.

26. Tattersall, "Dual Origin of Modern Humanity," 83.

27. K. R. Gibson, "Evolution of Human Intelligence: The Roles of Brain Size and Mental Construction," *Brain, Behavior and Evolution* 59, nos. 1–2 (2002): 10–20.

28. N. Humphrey, "Cave Art, Autism, and the Evolution of the Human Mind," *Cambridge Archaeological Journal* 8, no. 2 (1998): 165–191.

29. Tattersall, Ian, *Monkey in the Mirror*.

30. Wolfram Grajetzki, *Burial Customs in Ancient Egypt* (London: Duckworth, 2003); A. Rosalie David, *The Ancient Egyptians: Beliefs and Practices*, rev. and expanded ed., Sussex Library of Religious Beliefs and Practices (Portland, Ore.: Sussex Academic, 1998).

31. David, *Ancient Egyptians*, 22.

32. Grajetzki, *Burial Customs in Ancient Egypt*, 5.

33. David, *Ancient Egyptians*, 18.

34. Ibid., 79.

35. Stephen Quirke, *Ancient Egyptian Religion* (London: British Museum Press, 1992).

36. David, *Ancient Egyptians*, 57.

37. Richard W. Wrangham and Dale Peterson, *Demonic Males: Apes and the Origins of Human Violence* (Boston: Houghton Mifflin, 1996).

38. Nicholas Wade, *The Faith Instinct: How Religion Evolved and Why It Endures* (New York: Penguin, 2009).

39. Wrangham and Peterson, *Demonic Males*, 77.

40. Ibid., 67–68.

41. David Sloan Wilson, *Darwin's Cathedral: Evolution, Religion, and the Nature of Society* (Chicago: University of Chicago Press, 2002).

42. Charles Darwin, *The Descent of Man, and Selection in Relation to Sex* (New York: Appleton, 1871).

43. S. Bowles, "Did Warfare Among Ancestral Hunter-Gatherers Affect the Evolution of Human Social Behaviors?" *Science* 324 (2009): 1293–1298.

44. d'Errico et al., "Archaeological Evidence for the Emergence of Language, Symbolism, and Music," 40–42.

45. Wade, *Faith Instinct*, 100–111.

46. Ibid., 80.

47. Ibid., 88.

48. F. L. W. Ratnieks and T. Wenseleers, "Altruism in Insect Societies and Beyond: Voluntary or Enforced?" *Trends in Ecology and Evolution* 23, no. 1 (2008): 45–52.

49. P. M. Johns et al., "Nonrelatives Inherit Colony Resources in a Primitive Termite," *Proceedings of the National Academy of Sciences of the United States of America* 106, no. 41 (2009): 17452–17456.

50. Ibid., 17453–17454.

51. D. S. Wilson and E. O. Wilson, "Rethinking the Theoretical Foundation of Sociobiology," *Quarterly Review of Biology* 82, no. 4 (2007): 327–348.

52. E. O. Wilson and B. Holldobler, "Eusociality: Origin and Consequences," *Proceedings of the National Academy of Sciences of the United States of America* 102, no. 38 (2005): 13367–13371.

53. Wade, *Faith Instinct*, 52.

54. The New Jerusalem Bible (Garden City, N.Y.: Doubleday, 1985).

55. Pascal Boyer, *Religion Explained: The Evolutionary Origins of Religious Thought* (New York: Basic Books, 2001).

56. Ibid., 242–243.

57. Wade, *Faith Instinct*, 14.

58. Karen Armstrong, *Muhammad: A Biography of the Prophet*, 1st U.S. ed. (San Francisco: HarperSanFrancisco, 1992).

59. Edward O. Wilson, *On Human Nature* (Cambridge, Mass.: Harvard University Press, 1978).

60. Wrangham and Peterson, *Demonic Males*, 24.

61. Jane Goodall, *The Chimpanzees of Gombe: Patterns of Behavior* (Cambridge, Mass.: Belknap, 1986).

62. Wrangham and Peterson, *Demonic Males*, 16–21.

63. Ibid., 152.

9. Energy Flows

1. Peter H. Raven, Ray Franklin Evert, and Susan E. Eichhorn, *Biology of Plants*, 7th ed. (New York: Freeman, 2005).

2. Eric Chaisson, *Cosmic Evolution: The Rise of Complexity in Nature* (Cambridge, Mass.: Harvard University Press, 2001).

3. Ibid., 61; Eric D. Schneider and Dorion Sagan, *Into the Cool: Energy Flow, Thermodynamics, and Life* (Chicago: University of Chicago Press, 2005).

4. Schneider and Sagan, *Into the Cool*, 135–136.

5. Eric Chaisson, *The Life Era: Cosmic Selection and Conscious Evolution*, Evolutionary Synthesis Series (New York: Atlantic Monthly Press, 1987).

6. Schneider and Sagan, *Into the Cool*, 131–132.

7. I. Prigogine, *From Being to Becoming: Time and Complexity in the Physical Sciences* (San Francisco: Freeman, 1980).

8. E. J. Chaisson, "A Unifying Concept for Astrobiology," *International Journal of Astrobiology* 2, no. 2 (2003): 91–101.

9. Neil A. Campbell and Jane B. Reece, *Biology*, 8th ed. (San Francisco: Pearson Benjamin Cummings, 2009).

10. Juan G. Roederer, *Information and Its Role in Nature*, Frontiers Collection (Berlin: Springer, 2005).

11. D. Layzer, "Arrow of Time," *Scientific American* 233, no. 6 (1975): 56–69.

12. Chaisson, *Cosmic Evolution*, 141.

13. E. J. Chaisson, "Energy Rate Density as a Complexity Metric and Evolutionary Driver," *Complexity* 16, no. 3 (2011): 27–40.

14. E. J. Chaisson, "Energy Rate Density II: Probing Further a New Complexity Metric," *Complexity* (in press; early view published online at doi:10.1002/cplx .20373).

15. N. Lane and W. Martin, "The Energetics of Genome Complexity," *Nature* 467, no. 7318 (2010): 929–934.

16. Brian Keith McNab, *The Physiological Ecology of Vertebrates: A View from Energetics* (Ithaca: Cornell University Press, 2002).

17. Anthony K. Lee and Andrew Cockburn, *Evolutionary Ecology of Marsupials*, Monographs on Marsupial Biology (Cambridge: Cambridge University Press, 1985).

18. R. I. Dunbar and S. Shultz, "Understanding Primate Brain Evolution," *Philosophical Transactions of the Royal Society B—Biological Sciences* 362, no. 1480 (2007): 649–658.

19. K. Isler and C. P. van Schaik, "The Expensive Brain: A Framework for Explaining Evolutionary Changes in Brain Size," *Journal of Human Evolution* (2009): 392–400.

20. L. C. Aiello and J. C. K. Wells, "Energetics and the Evolution of the Genus Homo," *Annual Review of Anthropology* 31 (2002): 323–338.

21. David J. Randall et al., *Eckert Animal Physiology: Mechanisms and Adaptations*, 5th ed. (New York: Freeman, 2002).

22. B. K. McNab and J. F. Eisenberg. "Brain Size and Its Relation to the Rate of Metabolism in Mammals," *American Naturalist* 133, no. 2 (1989): 157–167.

23. K. Isler and C. van Schaik, "Costs of Encephalization: The Energy Trade-Off Hypothesis Tested on Birds," *Journal of Human Evolution* 51, no. 3 (2006): 228–243.

24. K. Isler et al., "Endocranial Volumes of Primate Species: Scaling Analyses Using a Comprehensive and Reliable Data Set," *Journal of Human Evolution* 55, no. 6 (2008): 967–978.

25. R. D. Martin, "Scaling of the Mammalian Brain: The Maternal Energy Hypothesis," *News in Physiological Sciences* 11 (1996): 149–156.

26. Based on a regression line for mammals in general. John Frederick Eisenberg, *The Mammalian Radiations: An Analysis of Trends in Evolution, Adaptation, and Behavior* (Chicago: University of Chicago Press, 1981).

27. Aiello and Wells, "Energetics," 331.

28. T. M. Preuss et al., "Human Brain Evolution: Insights from Microarrays," *Nature Reviews Genetics* 5, no. 11 (2004): 850–860.

29. A. Varki, D. H. Geschwind, and E. E. Eichler, "Explaining Human Uniqueness: Genome Interactions with Environment, Behaviour and Culture," *Nature Reviews Genetics* 9, no. 10 (2008): 749–763; M. Uddin et al., "Sister Grouping of Chimpanzees and Humans as Revealed by Genome-Wide Phylogenetic Analysis of Brain Gene Expression Profiles," *Proceedings of the National Academy of Sciences of the United States of America* 101, no. 9 (2004): 2957–2962.

30. W. R. Leonard, J. J. Snodgrass, and M. L. Robertson, "Effects of Brain Evolution on Human Nutrition and Metabolism," *Annual Review of Nutrition* 27 (2007): 311–327.

31. Ibid., 318.

32. R. N. Carmody and R. W. Wrangham, "The Energetic Significance of Cooking," *Journal of Human Evolution* (2009): 379–391.

33. L. C. Aiello and P. Wheeler, "The Expensive-Tissue Hypothesis—The Brain and the Digestive-System in Human and Primate Evolution," *Current Anthropology* 36, no. 2 (1995): 199–221.

34. Leonard, Snodgrass, and Robertson, "Effects of Brain Evolution," 320.

35. Aiello and Wells, "Energetics," 328.

36. William K. Purves, *Life, the Science of Biology*, 5th ed. (Sunderland, Mass.: Sinauer, 1998).

37. R. A. Berner, J. M. Vandenbrooks, and P. D. Ward, "Evolution: Oxygen and Evolution," *Science* 316, no. 5824 (2007): 557–558.

38. M. J. Benton, "Diversification and Extinction in the History of Life," *Science* 268, no. 5207 (1995): 52–58; M. J. Benton and B. C. Emerson, "How Did Life Become So Diverse? The Dynamics of Diversification According to the Fossil Record and Molecular Phylogenetics," *Palaeontology* 50 (2007): 23–40.

39. Geerat J. Vermeij, *Nature: An Economic History* (Princeton: Princeton University Press, 2004).

40. Ibid., 276.

41. R. K. Bambach, "Seafood Through Time—Changes in Biomass, Energetics, and Productivity in the Marine Ecosystem," *Paleobiology* 19, no. 3 (1993): 372–397.

42. Ibid., 376–378.

43. Ibid., 380.

44. Vermeij, *Nature*, 280.

45. R. K. Bambach, A. M. Bush, and D. H. Erwin, "Autecology and the Filling of Ecospace: Key Metazoan Radiations," *Palaeontology* 50 (2007): 1–22.

46. Ibid., 12–17.

47. S. Sahney, M. J. Benton, and P. A. Ferry, "Links Between Global Taxonomic Diversity, Ecological Diversity, and the Expansion of Vertebrates on Land," *Biology Letters* 6, no. 4 (2010): 544–547.

48. Edward O. Wilson, *The Diversity of Life*, Questions of Science (Cambridge, Mass.: Belknap, 1992), 187.

49. Vermeij, *Nature*, 10.

50. Ibid., 132.

51. McNab, *Physiological Ecology of Vertebrates*, 317.

52. Campbell and Reece, *Biology*, 1206.

53. L. N. Gillman and S. D. Wright, "The Influence of Productivity on the Species Richness of Plants: A Critical Assessment," *Ecology* 87, no. 5 (2006): 1234–1243.

54. P. Flombaum and O. E. Sala, "Higher Effect of Plant Species Diversity on Productivity in Natural Than Artificial Ecosystems," *Proceedings of the National Academy of Sciences of the United States of America* 105, no. 16 (2008): 6087–6090.

55. L. Johnson, "Macroecology: The Organizing Forces," *Biotechnology Progress* 22, no. 1 (2006): 156–166.

56. Stephen Jay Gould, *Full House: The Spread of Excellence from Plato to Darwin* (New York: Harmony, 1996).

57. J. D. Marcot and D. W. McShea, "Increasing Hierarchical Complexity Throughout the History of Life: Phylogenetic Tests of Trend Mechanisms," *Paleobiology* 33, no. 2 (2007): 182–200.

58. A. Kleidon, "Nonequilibrium Thermodynamics and Maximum Entropy Production in the Earth System," *Naturwissenschaften* 96, no. 6 (2009): 653–677.

59. A. Kleidon, "Non-Equilibrium Thermodynamics, Maximum Entropy Production and Earth-System Evolution," *Philosophical Transactions of the Royal Society A—Mathematical Physical and Engineering Sciences* 368, no. 1910 (2010): 181–196.

60. D. W. Schwartzman and T. Volk, "Biotic Enhancement of Weathering and the Habitability of Earth," *Nature* 340, no. 6233 (1989): 457–460.

61. Eugene Pleasants Odum, *Fundamentals of Ecology*, 3rd ed. (Philadelphia: Saunders, 1971).

62. Ibid., 257.

63. Schneider and Sagan, *Into the Cool*, 225–232.

64. Ibid., 221–223.

65. Ibid., 232–234.

66. E. G. Leigh and G. J. Vermeij, "Does Natural Selection Organize Ecosystems for the Maintenance of High Productivity and Diversity?" *Philosophical Transactions of the Royal Society B—Biological Sciences* 357, no. 1421 (2002): 709–718.

67. Chaisson, "Unifying Concept for Astrobiology," 96.

68. Vaclav Smil, *Energy in Nature and Society: General Energetics of Complex Systems* (Cambridge, Mass.: MIT Press, 2008).

69. Vaclav Smil, *General Energetics: Energy in the Biosphere and Civilization*, Environmental Science and Technology (New York: Wiley, 1991).

70. Ibid., 172–175.

71. Vaclav Smil, *Energies: An Illustrated Guide to the Biosphere and Civilization* (Cambridge, Mass.: MIT Press, 1999).

72. Smil, *General Energetics*, 307.

73. K. P. Gallagher, "Economic Globalization and the Environment," *Annual Review of Environment and Resources* 34 (2009): 279–304.

74. M. Chamon and M. Kremer, "Economic Transformation, Population Growth and the Long-Run World Income Distribution," *Journal of International Economics* 79, no. 1 (2009): 20–30.

75. Peter N. Stearns, *Globalization in World History* (New York: Routledge, 2009).

76. "Internet Usage Statistics," http://www.internetworldstats.com/stats.htm (accessed June 6, 2011).

77. "Global Swap Shops," *Economist*, January 30–February 5, 2010, p. 5.

78. "Facebook Attracts 700 Million Users Globally," *The Hindu*, http://www.thehindu.com/sci-tech/internet/article2074010.ece (accessed June 6, 2011).

79. Smil, *Energies*, 198.

80. Ibid., 200.

81. Schneider and Sagan, *Into the Cool*, 279.

82. Ibid., 141.

83. T. Agmon and A. Messica, "Financial Foreign Direct Investment: The Role of Private Equity Investments in the Globalization of Firms from Emerging Markets," *Management International Review* 49, no. 1 (2009): 11–25.

84. Stearns, *Globalization in World History*, 144.

10. The Origin of Life

1. P. Thaddeus, "The Prebiotic Molecules Observed in the Interstellar Gas," *Philosophical Transactions of the Royal Society B—Biological Sciences* 361, no. 1474 (2006): 1681–1687.

2. S. Kwok, "Delivery of Complex Organic Compounds from Planetary Nebulae to the Solar System," *International Journal of Astrobiology* 8, no. 3 (2009): 161–167.

3. H. Busemann et al., "Interstellar Chemistry Recorded in Organic Matter from Primitive Meteorites," *Science* 312, no. 5774 (2006): 727–730.

4. S. Pizzarello and E. Shock, "The Organic Composition of Carbonaceous Meteorites: The Evolutionary Story Ahead of Biochemistry," *Cold Spring Harbor Perspectives in Biology* 2, no. 3 (2010): a002105.

5. J. Jortner, "Conditions for the Emergence of Life on the Early Earth: Summary and Reflections," *Philosophical Transactions of the Royal Society B—Biological Sciences* 361, no. 1474 (2006): 1877–1891.

6. J. P. Ferris, "Montmorillonite-Catalysed Formation of RNA Oligomers: The Possible Role of Catalysis in the Origins of Life," *Philosophical Transactions of the Royal Society B—Biological Sciences* 361 (2006): 1777–1786.

7. N. Lane, J. F. Allen, and W. Martin, "How Did LUCA Make a Living? Chemiosmosis in the Origin of Life," *BioEssays* 32, no. 4 (2010): 271–280; W. K. Mat, H. Xue, and J. T. F. Wong, "The Genomics of LUCA," *Frontiers in Bioscience* 13 (2008): 5605–5613.

8. W. Martin et al., "Hydrothermal Vents and the Origin of Life," *Nature Reviews Microbiology* 6, no. 11 (2008): 805–814.

9. H. J. Cleaves et al., "A Reassessment of Prebiotic Organic Synthesis in Neutral Planetary Atmospheres," *Origins of Life and Evolution of Biospheres* 38, no. 2 (2008): 105–115.

10. O. Abramov and S. J. Mojzsis, "Microbial Habitability of the Hadean Earth During the Late Heavy Bombardment," *Nature* 459, no. 7245 (2009): 419–422.

11. T. M. Harrison, "The Hadean Crust: Evidence from >4 Ga Zircons," *Annual Review of Earth and Planetary Sciences* 37 (2009): 479–505; S. J. Mojzsis, "Early Earth Leftover Lithosphere," *Nature Geoscience* 3, no. 3 (2010): 148–149.

12. Pizzarello and Shock, "Organic Composition of Carbonaceous Meteorites," 15.

13. M. J. Van Kranendonk, "Volcanic Degassing, Hydrothermal Circulation and the Flourishing of Early Life on Earth: A Review of the Evidence from C. 3490–3240 Ma Rocks of the Pilbara Supergroup, Pilbara Craton, Western Australia," *Earth-Science Reviews* 74, no. 3-4 (2006): 197–240.

14. G. Ryder, "Bombardment of the Hadean Earth: Wholesome or Deleterious?" *Astrobiology* 3, no. 1 (2003): 3–6.

15. C. S. Cockell, "The Origin and Emergence of Life Under Impact Bombardment," *Philosophical Transactions of the Royal Society B—Biological Sciences* 361, no. 1474 (2006): 1845–1855; discussion 56.

16. S. Pizzarello, "The Chemistry of Life's Origin: A Carbonaceous Meteorite Perspective," *Accounts of Chemical Research* 39, no. 4 (2006): 231–237; Pizzarello and Shock, "Organic Composition of Carbonaceous Meteorites," 4.

17. G. M. M. Caro et al., "Amino Acids from Ultraviolet Irradiation of Interstellar Ice Analogues," *Nature* 416, no. 6879 (2002): 403–406.

18. M. M. Hanczyc, S. M. Fujikawa, and J. W. Szostak, "Experimental Models of Primitive Cellular Compartments: Encapsulation, Growth, and Division," *Science* 302, no. 5645 (2003): 618–622.

19. D. W. Deamer and J. P. Dworkin, "Chemistry and Physics of Primitive Membranes," *Prebiotic Chemistry: From Simple Amphiphiles to Protocell Models* 259 (2005): 1–27.

20. D. Segre et al., "The Lipid World," *Origins of Life and Evolution of the Biosphere* 31, nos. 1–2 (2001): 119–145.

21. D. Deamer and A. L. Weber, "Bioenergetics and Life's Origins," *Cold Spring Harbor Perspectives in Biology* 2, no. 2 (2010): 1–16.

22. Deamer and Dworkin, "Chemistry and Physics of Primitive Membranes," 12.

23. E. G. Nisbet and N. H. Sleep, "The Habitat and Nature of Early Life," *Nature* 409, no. 6823 (2001): 1083–1091.

24. R. Sanchez, J. Ferris, and L. E. Orgel, "Conditions for Purine Synthesis: Did Prebiotic Synthesis Occur at Low Temperatures?" *Science* 153, no. 731 (1966): 72–73.

25. Ferris, "Montmorillonite-Catalysed Formation of RNA Oligomers," 1778–1779.

26. M. P. Robertson and S. L. Miller, "An Efficient Prebiotic Synthesis of Cytosine and Uracil," *Nature* 375, no. 6534 (1995): 772–774.

27. G. Cooper et al., "Carbonaceous Meteorites as a Source of Sugar-Related Organic Compounds for the Early Earth," *Nature* 414, no. 6866 (2001): 879–883.

28. W. Nitschke and M. J. Russell, "Hydrothermal Focusing of Chemical and Chemiosmotic Energy, Supported by Delivery of Catalytic Fe, Ni, Mo/W, Co, S and Se, Forced Life to Emerge," *Journal of Molecular Evolution* 69, no. 5 (2009): 481–496.

29. M. W. Powner, B. Gerland, and J. D. Sutherland, "Synthesis of Activated Pyrimidine Ribonucleotides in Prebiotically Plausible Conditions," *Nature* 459, no. 7244 (2009): 239–242.

30. Cooper et al., "Carbonaceous Meteorites as a Source of Sugar-Related Organic Compounds," 881–882.

31. Deamer and Weber, "Bioenergetics and Life's Origins," 5.

32. P. Baaske et al., "Extreme Accumulation of Nucleotides in Simulated Hydrothermal Pore Systems," *Proceedings of the National Academy of Sciences of the United States of America* 104, no. 22 (2007): 9346–9351.

33. Ibid., 9347.

34. Ibid., 9348–9350.

35. S. E. McGlynn et al., "Hydrogenase Cluster Biosynthesis: Organometallic Chemistry Nature's Way," *Dalton Trans*, no. 22 (2009): 4274–4285.

36. S. D. Copley, E. Smith, and H. J. Morowitz, "The Origin of the RNA World: Co-Evolution of Genes and Metabolism," *Bioorganic Chemistry* 35, no. 6 (2007): 430–443.

37. T. A. Lincoln and G. F. Joyce, "Self-Sustained Replication of an RNA Enzyme," *Science* 323, no. 5918 (2009): 1229–1232.

38. M. Yarus, J. J. Widmann, and R. Knight, "RNA-Amino Acid Binding: A Stereochemical Era for the Genetic Code," *Journal of Molecular Evolution* 69, no. 5 (2009): 406–429.

39. S. Rajamani et al., "Lipid-Assisted Synthesis of RNA-Like Polymers from Mononucleotides," *Origins of Life and Evolution of the Biosphere* 38, no. 1 (2008): 57–74.

40. M. Tessera, "Life Began When Evolution Began: A Lipidic Vesicle-Based Scenario," *Origins of Life and Evolution of the Biosphere* 39, no. 6 (2009): 559–564.

41. Lane, Allen, and Martin, "How Did LUCA Make a Living?" 3.

42. David E. Sadava et al., *Life, the Science of Biology*, 8th ed. (Sunderland, Mass.: Sinauer, 2008).

43. A. Serrano et al., "H+-Ppases: Yesterday, Today and Tomorrow," *IUBMB Life* 59, no. 2 (2007): 76–83.

44. M. J. Russell and A. J. Hall, "The Emergence of Life from Iron Monosulphide Bubbles at a Submarine Hydrothermal Redox and Ph Front," *Journal of the Geological Society of London* 154, no. 3 (1997): 377–402.

45. V. Srinivasan and H. J. Morowitz, "The Canonical Network of Autotrophic Intermediary Metabolism: Minimal Metabolome of a Reductive Chemoautotroph," *Biological Bulletin* 216, no. 2 (2009): 126–130.

46. E. Smith and H. J. Morowitz, "Universality in Intermediary Metabolism," *Proceedings of the National Academy of Sciences of the United States of America* 101, no. 36 (2004): 13168–13173.

47. M. Illangasekare and M. Yarus, "A Tiny RNA that Catalyzes Both Aminoacyl-RNA and Peptidyl-RNA Synthesis," *RNA* 5, no. 11 (1999): 1482–1489.

48. E. V. Koonin and W. Martin, "On the Origin of Genomes and Cells Within Inorganic Compartments," *Trends in Genetics* 21, no. 12 (2005): 647–654.

49. Harrison, "Hadean Crust," 489.

50. Van Kranendonk, "Volcanic Degassing, Hydrothermal Circulation and the Flourishing of Early Life on Earth," 232.

51. Koonin and Martin, "On the Origin of Genomes and Cells," 650.

52. T. Dagan, Y. Artzy-Randrup, and W. Martin, "Modular Networks and Cumulative Impact of Lateral Transfer in Prokaryote Genome Evolution," *Proceedings of the National Academy of Sciences of the United States of America* 105, no. 29 (2008): 10039–10044.

53. V. Srinivasan and H. J. Morowitz, "Analysis of the Intermediary Metabolism of a Reductive Chemoautotroph," *Biological Bulletin* 217, no. 3 (2009): 222–232.

54. Horst Rauchfuss, *Chemical Evolution and the Origin of Life* (New York: Springer, 2008).

55. Van Kranendonk, "Volcanic Degassing, Hydrothermal Circulation and the Flourishing of Early Life on Earth," 235.

56. F. U. Battistuzzi, A. Feijao, and S. B. Hedges, "A Genomic Timescale of Prokaryote Evolution: Insights into the Origin of Methanogenesis, Phototrophy, and the Colonization of Land," *BMC Evolutionary Biology* 4, no. 44 (2004).

57. Harrison, "Hadean Crust," 496.

58. Ibid., 489.

59. Abramov and Mojzsis, "Microbial Habitability of the Hadean Earth," 421.

60. Srinivasan and Morowitz, "Canonical Network of Autotrophic Intermediary Metabolism," 127.

61. K. Kashefi and D. R. Lovley, "Extending the Upper Temperature Limit for Life," *Science* 301, no. 5635 (2003): 934.

62. L. J. Rothschild and R. L. Mancinelli, "Life in Extreme Environments," *Nature* 409, no. 6823 (2001): 1092–1101.

63. L. H. Lin et al., "Long-Term Sustainability of a High-Energy, Low-Diversity Crustal Biome," *Science* 314, no. 5798 (2006): 479–482.

64. L. J. Rothschild, "Earth Science Life Battered but Unbowed," *Nature* 459, no. 7245 (2009): 335–336.

65. "NASA Spots Surprising Shrimp Beneath Antarctic Ice," http://www.nasa.gov/topics/earth/features/antarctic-shrimp.html (accessed June 7, 2011).

66. S. Conway Morris, *Life's Solution: Inevitable Humans in a Lonely Universe* (Cambridge: Cambridge University Press, 2003).

67. Christian De Duve, *Vital Dust: Life as a Cosmic Imperative* (New York: Basic, 1995), 121.

11. The Prospects for Habitable Worlds

1. Fred J. Ciesla and C. P. Dullemond, "Evolution of Protoplanetary Disk Structures," in *Protoplanetary Dust: Astrochemical and Cosmochemical Perspectives*, ed. Dániel Apai and D. S. Lauretta (Cambridge: Cambridge University Press, 2010), 66–96; A. V. Krivov, "Debris Disks: Seeing Dust, Thinking of Planetesimals and Planets," *Research in Astronomy and Astrophysics* 10, no. 5 (2010): 383–414.

2. Eric Chaisson and S. McMillan, *Astronomy Today*, 5th ed. (Upper Saddle River, N.J.: Pearson/Prentice Hall, 2005).

3. Jeffrey O. Bennett, *The Cosmic Perspective*, 4th ed. (San Francisco: Pearson/Addison-Wesley, 2007).

4. Chaisson and McMillan, *Astronomy Today*, 289.

5. Bennett, *Cosmic Perspective*, 4th ed., 593.

6. R. Jayawardhana and V. D. Ivanov, "Spectroscopy of Young Planetary Mass Candidates with Disks," *Astrophysical Journal* 647, no. 2 (2006): L167–L170.

7. Z. X. Wang, D. Chakrabarty, and D. L. Kaplan, "A Debris Disk Around an Isolated Young Neutron Star," *Nature* 440, no. 7085 (2006): 772–775.

8. H. J. Habing et al., "Disappearance of Stellar Debris Disks Around Main-Sequence Stars After 400 Million Years," *Nature* 401, no. 6752 (1999): 456–458.

9. G. F. Benedict et al., "The Extrasolar Planet Is an Element of Eridani B: Orbit and Mass," *Astronomical Journal* 132, no. 5 (2006): 2206–2218.

10. Krivov, "Debris Disks," 405–407.

11. Bennett, *Cosmic Perspective*, 4th ed., 232.

12. Eric Chaisson and S. McMillan, *Astronomy: A Beginner's Guide to the Universe*, 5th ed. (Upper Saddle River, N.J.: Pearson/Prentice Hall, 2007).

13. Bennett, *Cosmic Perspective*, 4th ed., 234.

14. James F. Kasting, *How to Find a Habitable Planet* (Princeton: Princeton University Press, 2010).

15. E. B. Ford and K. D. Colon, "Characterizing the Eccentricities of Transiting Extrasolar Planets with Kepler and Corot," *Transiting Planets, Proceedings* 4, no. 253 (2009): 111–119.

16. Bennett, *Cosmic Perspective*, 4th ed., 543.

17. Ibid., 517.

18. Iain Gilmour, Mark A. Sephton, and Andrew Conway, *An Introduction to Astrobiology* (Milton Keynes, U.K.: Open University; New York: Cambridge University Press, 2004).

19. Jeffrey Bennett, *The Cosmic Perspective*, 6th ed. (Boston: Addison-Wesley, 2010).

20. Kasting, *How to Find a Habitable Planet*, 177–179.

21. P. Ulmschneider, *Intelligent Life in the Universe: From Common Origins to the Future of Humanity*, Advances in Astrobiology and Biogeophysics (New York: Springer-Verlag, 2003).

22. Joseph Silk, *The Infinite Cosmos: Questions from the Frontiers of Cosmology* (Oxford: Oxford University Press, 2006).

23. Bennett, *Cosmic Perspective*, 4th ed., 598.

24. Kevin W. Plaxco and Michael Gross, *Astrobiology: A Brief Introduction* (Baltimore: Johns Hopkins University Press, 2006); Bennett, *Cosmic Perspective*, 6th ed., 549.

25. Neil deGrasse Tyson and Donald Goldsmith, *Origins: Fourteen Billion Years of Cosmic Evolution* (New York: Norton, 2004).

26. Bennett, *Cosmic Perspective*, 6th ed., 525.

27. M. Trenti, "Astronomy: Galaxy Sets Distance Mark," *Nature* 467, no. 7318 (2010): 924–925.

28. C. J. Lada, "Stellar Multiplicity and the Initial Mass Function: Most Stars Are Single," *Astrophysical Journal* 640, no. 1 (2006): L63–L66; M. J. Heath et al., "Habitability of Planets Around Red Dwarf Stars," *Origins of Life and Evolution of the Biosphere* 29, no. 4 (1999): 405–424.

29. F. C. Adams, P. Bodenheimer, and G. Laughlin, "M Dwarfs: Planet Formation and Long Term Evolution," *Astronomische Nachrichten* 326, no. 10 (2005): 913–919.

30. M. Werner et al., "First Fruits of the Spitzer Space Telescope: Galactic and Solar System Studies," *Annual Review of Astronomy and Astrophysics* 44 (2006): 269–321.

31. G. M. Kennedy, S. J. Kenyon, and B. C. Bromley, "Planet Formation Around Low-Mass Stars: The Moving Snow Line and Super-Earths," *Astrophysical Journal* 650, no. 2 (2006): L139–L142.

32. B. S. Gaudi et al., "Discovery of a Jupiter/Saturn Analog with Gravitational Microlensing," *Science* 319, no. 5865 (2008): 927–930.

33. L. Szucs et al., "Stellar-Mass-Dependent Disk Structure in Coeval Planet-Forming Disks," *Astrophysical Journal* 720, no. 2 (2010): 1668–1673.

34. J. M. Griessmeier et al., "Cosmic Ray Impact on Extrasolar Earth-Like Planets in Close-in Habitable Zones," *Astrobiology* 5, no. 5 (2005): 587–603.

35. Heath et al., "Habitability of Planets Around Red Dwarf Stars," 405–424.

36. Bennett, *Cosmic Perspective*, 4th ed., 551.

37. Ibid., 272.

38. Heath et al., "Habitability of Planets Around Red Dwarf Stars," 407.

39. Bennett, *Cosmic Perspective*, 6th ed., 203.

40. J. M. Griessmeier et al., "On the Protection of Extrasolar Earth-Like Planets Around K/M Stars Against Galactic Cosmic Rays," *Icarus* 199, no. 2 (2009): 526–535.

41. Jeffrey O. Bennett, *Beyond UFOs: The Search for Extraterrestrial Life and Its Astonishing Implications for Our Future* (Princeton: Princeton University Press, 2008).

42. L. Kaltenegger, "Characterizing Habitable Exomoons," *Astrophysical Journal Letters* 712, no. 2 (2010): L125–L130.

43. K. Todorov, K. L. Luhman, and K. K. McLeod, "Discovery of a Planetary-Mass Companion to a Brown Dwarf in Taurus," *Astrophysical Journal Letters* 714, no. 1 (2010): L84–L88.

44. D. Raghavan et al., "Two Suns in the Sky: Stellar Multiplicity in Exoplanet Systems," *Astrophysical Journal* 646, no. 1 (2006): 523–542.

45. L. A. Saleh and F. A. Rasio, "The Stability and Dynamics of Planets in Tight Binary Systems," *Astrophysical Journal* 694, no. 2 (2009): 1566–1576; E. V. Quintana and J. J. Lissauer, "Terrestrial Planet Formation Surrounding Close Binary Stars," *Icarus* 185, no. 1 (2006): 1–20.

46. N. Haghighipour, "Binary Stars with Habitable Planets," *American Scientist* 96, no. 4 (2008): 294–301.

47. Quintana and Lissauer, "Terrestrial Planet Formation," 2.

48. Lada, "Stellar Multiplicity and the Initial Mass Function," L63–L64.

49. Silk, *Infinite Cosmos*, 57.

50. Ibid., 26.

51. C. H. Lineweaver, Y. Fenner, and B. K. Gibson, "The Galactic Habitable Zone and the Age Distribution of Complex Life in the Milky Way," *Science* 303, no. 5654 (2004): 59–62.

52. N. Prantzos, "On the 'Galactic Habitable Zone,'" *Space Science Reviews* 135, nos. 1–4 (2008): 313–322.

53. A. Heavens, "The Star-Formation History of the Universe," *American Scientist* 93, no. 1 (2005): 36–41.

54. L. J. Tacconi et al., "High Molecular Gas Fractions in Normal Massive Star-Forming Galaxies in the Young Universe," *Nature* 463, no. 7282 (2010): 781–784.

55. Bennett, *Cosmic Perspective*, 6th ed., 590–591.

56. A. Dekel et al., "Cold Streams in Early Massive Hot Haloes as the Main Mode of Galaxy Formation," *Nature* 457, no. 7228 (2009): 451–454.

57. Bennett, *Cosmic Perspective*, 6th ed., 592–608.

58. A. Cattaneo et al., "The Role of Black Holes in Galaxy Formation and Evolution," *Nature* 460, no. 7252 (2009): 213–219.

59. Bennett, *Cosmic Perspective*, 6th ed., 632–633.

60. Michael A. Seeds, *Horizons: Exploring the Universe*, 9th ed. (Belmont, Calif.: Thomson Brooks/Cole, 2006).

61. A. Antunes and J. Wallin, "Dynamical Parameters for Am 0644-741," *Astrophysical Journal* 670, no. 1 (2007): 261–268; C. Struck, "Applying the Analytic Theory of Colliding Ring Galaxies," *Monthly Notices of the Royal Astronomical Society* 403, no. 3 (2010): 1516–1530.

62. D. A. Fischer and J. Valenti, "The Planet-Metallicity Correlation," *Astrophysical Journal* 622, no. 2 (2005): 1102–1117.

63. Ibid., 1116.

64. Bennett, *Cosmic Perspective*, 6th ed., 543–549.

65. Hyron Spinrad, *Galaxy Formation and Evolution*, Springer-Praxis Books in Astrophysics and Astronomy (Berlin: Springer, in association with Praxis, Chichester, U.K., 2005).

66. Seeds, *Horizons*, 266.

67. J. Setiawan et al., "A Giant Planet Around a Metal-Poor Star of Extragalactic Origin," *Science* 330, no. 6011 (2010): 1642–1644.

68. B. Ercolano and C. J. Clarke, "Metallicity, Planet Formation and Disc Lifetimes," *Monthly Notices of the Royal Astronomical Society* 402, no. 4 (2010): 2735–2743.

69. Eric Gaidos and Franck Selsis, "From Protoplanets to Protolife: The Emergence and Maintenance of Life," in *Protostars and Planets V*, ed. Bo Reipurth, D. Jewitt, and Klaus Keil (Tucson: University of Arizona Press, in collaboration with Lunar and Planetary Institute, Houston, 2007), 929–944.

70. Fischer and Valenti, "Planet-Metallicity Correlation," 1116.

71. Peter Douglas Ward and Donald Brownlee, *Rare Earth: Why Complex Life Is Uncommon in the Universe* (New York: Copernicus, 2000).

72. Kasting, *How to Find a Habitable Planet*, 162–163.

73. Ward and Brownlee, *Rare Earth*, 224–226.

74. S. Elser et al., "How Common Are Earth-Moon Planetary Systems?" *Icarus* (2011, in press).

75. Bennett, *Beyond UFOs*, 180.

76. Kasting, *How to Find a Habitable Planet*, 164–169.

77. Bennett, *Beyond UFOs*, 118–119.

78. Ulmschneider, *Intelligent Life in the Universe*, 64.

79. Bennett, *Beyond UFOs*, 118.

80. Ward and Brownlee, *Rare Earth*, 210.

81. Kasting, *How to Find a Habitable Planet*, 108–112.

82. Peter Douglas Ward, *Life as We Do Not Know It: The NASA Search for (and Synthesis of) Alien Life* (New York: Viking, 2005).

83. Kasting, *How to Find a Habitable Planet*, 145–148.

84. Bennett, *Cosmic Perspective*, 6th ed., 710.

85. Haghighipour, "Binary Stars with Habitable Planets," 299.

86. Ward and Brownlee, *Rare Earth*, 213.

87. Kasting, *How to Find a Habitable Planet*, 149.

88. Ibid., 151.

89. Bennett, *Beyond UFOs*, 151.

90. Ibid., 132.

91. Kasting, *How to Find a Habitable Planet*, 177–179.

92. Bennett, *Beyond UFOs*, 114.

93. Kasting, *How to Find a Habitable Planet*, 178.

94. Ibid., 118.

95. M. J. Van Kranendonk, "Volcanic Degassing, Hydrothermal Circulation and the Flourishing of Early Life on Earth: A Review of the Evidence from C. 3490–3240 Ma Rocks of the Pilbara Supergroup, Pilbara Craton, Western Australia," *Earth-Science Reviews* 74, no. 3-4 (2006): 197–240.

96. G. Ryder, "Bombardment of the Hadean Earth: Wholesome or Deleterious?" *Astrobiology* 3, no. 1 (2003): 3–6.

97. F. A. Corsetti, S. M. Awramik, and D. Pierce, "A Complex Microbiota from Snowball Earth Times: Microfossils from the Neoproterozoic Kingston Peak Formation, Death Valley, USA," *Proceedings of the National Academy of Sciences of the United States of America* 100, no. 8 (2003): 4399–4404.

98. K. Junge, H. Eicken, and J. W. Deming, "Bacterial Activity at −2 to −20 Degrees C in Arctic Wintertime Sea Ice," *Applied and Environmental Microbiology* 70, no. 1 (2004): 550–557.

99. Setiawan et al., "Giant Planet Around a Metal-Poor Star," 2.

100. D. Chivian et al., "Environmental Genomics Reveals a Single-Species Ecosystem Deep Within Earth," *Science* 322, no. 5899 (2008): 275–278.

101. L. J. Rothschild and R. L. Mancinelli, "Life in Extreme Environments," *Nature* 409, no. 6823 (2001): 1092–1101.

102. "Tardigrade," http://en.wikipedia.org/wiki/Tardigrade (accessed June 9, 2011).

103. G. Borgonie et al., "Nematoda from the Terrestrial Deep Subsurface of South Africa," *Nature* 474, no. 7349 (2011): 79–82.

104. Marc Kaufman, "'Worms from Hell' Unearth Possibilities for Extraterrestrial Life," *Washington Post*, June 2, 2011, http://www.ongo.com/v/1061484/-1/069C3A2682E62B52/discovery-of-worms-from-hell-deep-beneath-earths-surface-raises-new-questions (accessed June 2, 2011).

105. K. O. Stetter, "Hyperthermophiles in the History of Life," *Philosophical Transactions of the Royal Society B—Biological Sciences* 361, no. 1474 (2006): 1837–1842; discussion 42–43.

106. Rothschild and Mancinelli, "Life in Extreme Environments," 1097.

107. Helga Stan-Lotter, "Extremophiles, the Physicochemical Limits of Life (Growth and Survival)," in *Complete Course in Astrobiology*, ed. Gerda Horneck and Petra Rettberg (Weinheim, Germany: Wiley, 2007), 121–150.

108. "Lake Natron," http://en.wikipedia.org/wiki/Lake_Natron (accessed June 9, 2011).

109. Josep del Hoyo et al., *Handbook of the Birds of the World*, vol. 1 (Barcelona: Lynx Edicions, 1992).

110. Bernhard Grzimek et al., *Grzimek's Animal Life Encyclopedia*, 2nd ed., vol. 8 (Detroit: Gale, 2003).

111. "Antarctic Connection," http://www.antarcticconnection.com/antarctic/wildlife/penguins/emperor.shtml (accessed June 7, 2011).

112. Guillaume Dargaud, "Antarctic Penguins," http://www.gdargaud.net/Antartica/Penguins.html (accessed June 7, 2011).

113. Ulmschneider, *Intelligent Life in the Universe*, 74–75.

114. Kasting, *How to Find a Habitable Planet*, 10.

115. Ibid., 296.

116. J. P. Guo et al., "Probability Distribution of Terrestrial Planets in Habitable Zones Around Host Stars," *Astrophysics and Space Science* 323, no. 4 (2009): 367–373.

117. W. Lanouette, *Genius in the Shadows* (New York: Scribner, 1992), 257.

118. Chaisson and McMillan, *Astronomy Today*, 756.

119. Bennett, *Cosmic Perspective*, 4th ed., 728.

120. Tyson and Goldsmith, *Origins*, 285.

121. Kasting, *How to Find a Habitable Planet*, 241.

122. Ibid., 261–266.

123. Ibid., 279–281.

124. Griessmeier et al., "On the Protection of Extrasolar Earth-Like Planets," 533.

125. C. J. Grillmair et al., "Strong Water Absorption in the Dayside Emission Spectrum of the Planet Hd189733b," *Nature* 456, no. 7223 (2008): 767–769.

126. Kasting, *How to Find a Habitable Planet*, 234–238.

127. Ibid., 238.

128. "The Extrasolar Planets Encyclopaedia," http://exoplanet.eu (accessed June 7, 2011).

129. C. Mordasini et al., "Extrasolar Planet Population Synthesis II. Statistical Comparison with Observations," *Astronomy and Astrophysics* 501, no. 3 (2009): 1161–1184.

130. G. Laughlin, "A Dance of Extrasolar Planets," *Science* 330, no. 6000 (2010): 47–48.

131. M. Mewhinney, R. Hoover, and T. J. Perrotto, "NASA Finds Earth-size Planet Candidates in the Habitable Zone," http://www.nasa.gov/mission_pages/kepler/news/kepler_data_release.html (accessed June 7, 2011).

132. E. S. Reich, "Astronomy: Beyond the Stars," *Nature* 470, no. 7332 (2011): 24–26.

133. Mewhinney, "NASA Finds Earth-size Planet Candidates in the Habitable Zone."

134. Dennis Overbye, "Kepler Planet Hunter Finds 1,200 Possibilities," *New York Times*, February 2, 2011, http://www.nytimes.com/2011/02/03/science/03planet.html (accessed June 7, 2011); Franck Marchis, "A Landslide of Kepler Exoplanet Candidates," The SETI Institute, February 2, 2011, http://scienceblogs.com/SETI/2011/02/a_landslide_of_kepler_exoplane.php (accessed June 7, 2011).

12. The Apex of Nature

1. Jeffrey Bennett, *The Cosmic Perspective*, 6th ed. (Boston: Addison-Wesley, 2010), 532.

2. Carl Sagan, *The Dragons of Eden: Speculations on the Evolution of Human Intelligence* (New York: Random House, 1977).

3. S. Herculano-Houzel, "Brains Matter, Bodies Maybe Not: The Case for Examining Neuron Numbers Irrespective of Body Size," *Annals of the New York Academy of Sciences* 1225, no. 1 (2011): 191–199.

4. Michael Mallary, *Our Improbable Universe: A Physicist Considers How We Got Here* (New York: Thunder's Mouth, 2004); Holmes Rolston, *Three Big Bangs: Matter-Energy, Life, Mind* (New York: Columbia University Press, 2010).

5. Mallary, *Our Improbable Universe*, 18.

6. Joseph Silk, *The Infinite Cosmos: Questions from the Frontiers of Cosmology* (Oxford: Oxford University Press, 2006).

7. Mallary, *Our Improbable Universe*, 27.

8. Ibid., 5, 18.

9. Ibid., 7.

10. P. C. W. Davies, *Cosmic Jackpot: Why Our Universe Is Just Right for Life* (Boston: Houghton Mifflin, 2007).

11. Ibid., 149–150.

12. M. Tegmark, "Parallel Universes," *Scientific American* 288, no. 5 (2003): 40–51.

13. C. Romanzin et al., "Water Formation by Surface O-3 Hydrogenation," *Journal of Chemical Physics* 134, no. 8 (2011, in press).

14. F. Dulieu et al., "Experimental Evidence for Water Formation on Interstellar Dust Grains by Hydrogen and Oxygen Atoms," *Astronomy and Astrophysics* 512 (2010).

15. Kevin W. Plaxco and Michael Gross, *Astrobiology: A Brief Introduction* (Baltimore: Johns Hopkins University Press, 2006).

16. Ira Blei and George G. Odian, *General, Organic, and Biochemistry: Connecting Chemistry to Your Life*, 2nd ed. (New York: Freeman, 2006).

17. Jane B. Reece and Neil A. Campbell, *Biology*, 9th ed. (San Francisco: Pearson Benjamin Cummings, 2011).

18. D. C. Catling et al., "Why O_2 Is Required by Complex Life on Habitable Planets and the Concept of Planetary 'Oxygenation Time,'" *Astrobiology* 5, no. 3 (2005): 415–438.

19. Ibid., 430.

20. Davies, *Cosmic Jackpot*, 135.

21. Mallary, *Our Improbable Universe*, 54–56.

22. H. Oberhummer, A. Csoto, and H. Schlattl, "Stellar Production Rates of Carbon and Its Abundance in the Universe," *Science* 289, no. 5476 (2000): 88–90.

23. Mallary, *Our Improbable Universe*, 55.

24. Oberhummer, Csoto, and Schlattl, "Stellar Production Rates of Carbon," 88.

25. Mallary, *Our Improbable Universe*, 7.

26. S. Ekstrom et al., "Effects of the Variation of Fundamental Constants on Population Iii Stellar Evolution," *Astronomy and Astrophysics* 514 (2010).

27. E. Gaidos et al., "Beyond the Principle of Plentitude: A Review of Terrestrial Planet Habitability," *Astrobiology* 5, no. 2 (2005): 100–126.

28. Roger Penrose, *The Road to Reality: A Complete Guide to the Laws of the Universe*, 1st Vintage Books ed. (New York: Vintage, 2007), 706–715.

29. David Layzer, *Cosmogenesis: The Growth of Order in the Universe* (New York: Oxford University Press, 1990), 35–36.

30. Reece and Campbell, *Biology*, 529.

31. Tegmark, "Parallel Universes," 41.

32. Martin Gardner, "Multiverses and Blackberries," *Skeptical Inquirer* 25.5 (2001), http://www.csicop.org/si/show/multiverses_and_blackberries (accessed June 6, 2011).

33. Davies, *Cosmic Jackpot*, 81.

34. B. Greene, *The Hidden Reality: Parallel Universes and the Deep Laws of the Cosmos* (New York: Knopf, 2011).

35. Steven Weinberg, *Dreams of a Final Theory*, 1st Vintage Books ed. (New York: Vintage, 1994), 192.

36. Brian Greene, *The Elegant Universe: Superstrings, Hidden Dimensions, and the Quest for the Ultimate Theory* (New York: Norton, 1999).

37. Davies, *Cosmic Jackpot*, 154.

38. Quoted in Weinberg, *Dreams of a Final Theory*, 230.

39. Pierre Teilhard de Chardin, *The Phenomenon of Man* (New York: Harper, 1959).

40. Ibid., 165.

41. Edward O. Wilson, *On Human Nature* (Cambridge, Mass.: Harvard University Press, 1978).

42. Richard W. Wrangham and Dale Peterson, *Demonic Males: Apes and the Origins of Human Violence* (Boston: Houghton Mifflin, 1996).

43. Nicholas Wade, *The Faith Instinct: How Religion Evolved and Why It Endures* (New York: Penguin, 2009).

44. T. Bersaglieri et al., "Genetic Signatures of Strong Recent Positive Selection at the Lactase Gene." *American Journal of Human Genetics* 74, no. 6 (2004): 1111–1120.

45. Pascal Boyer, *Religion Explained: The Evolutionary Origins of Religious Thought* (New York: Basic, 2001).

46. A theology is presented in both prose and poetry in Raymond L. Neubauer, *Voyages into Transcendence* (Austin: Bay of Rainbows Press, 2002).

47. Teilhard de Chardin, *Phenomenon of Man*, 272.

Index